T0182118

Random Numbers and Computers

Ronald T. Kneusel

Random Numbers and Computers

 Springer

Ronald T. Kneusel
Thornton, CO, USA

ISBN 978-3-030-08516-2   ISBN 978-3-319-77697-2   (eBook)
https://doi.org/10.1007/978-3-319-77697-2

Printed on acid-free paper

This Springer imprint is published by the registered company Springer International Publishing AG part of Springer Nature.
The registered company address is: Gewerbestrasse 11, 6330 Cham, Switzerland

To my wife, Maria, for her love and her patience. It is the randomness of life that makes it most exciting, especially when shared.

Preface

This is a book about random numbers and computers. That is an ambiguous sentence, so let's be more precise: this is a book about pseudorandom numbers and computers. While closer to the mark, we are not quite there, so let's try to be still more precise: this is a book about algorithms capable of generating sequences of numbers that, according to a series of statistical tests, are suitably indistinguishable from number sequences generated by true random processes.

Let's break it down some more:

- Computers often need sequences of numbers where it is not possible, given n_i, to predict n_{i+1}. Ideally, n_{i+1} should not be predictable from any combination of the previous k numbers in the sequence (for any k).
- While true random processes exist, an assumption on our part we will not prove, it is not typically practical for computers to make use of random processes so we are left with algorithmic processes. Computers are good at following algorithms.
- This is a book about algorithms that can generate sequences of numbers approximating the requirements of a true random sequence. We call these algorithms *pseudorandom number generators* (PRNGs).
- So, then, in the end, this is a book about pseudorandom number generators: how they work, how to test them, and how to use them with computers.

The above implies that this is not a book about randomness *per se*. Concepts related to random sequences are discussed briefly in Chap. 1 but only as far as is necessary to motivate the introduction and discussion of pseudorandomness and eventually pseudorandom number generators.

The goal of the book is to equip the reader with enough background in pseudorandom number generators to understand how they work and how to demonstrate they are useful. We will do this by developing actual code, in C or Python, to make the algorithms concrete. Along the way we will run various experiments designed to make the code practical and to demonstrate key concepts from the discussion.

A Note About Terminology

The phrases "pseudorandom number" and "pseudorandom number generator" are somewhat tedious. Therefore, we will frequently succumb to the temptation to write "random number" or "random number generator" when we in fact mean "pseudorandom number" or "pseudorandom number generator." We will rely on context to clarify when these phrases are used loosely or precisely. We trust that the reader will understand and follow along without difficulty.

Who Should Read This Book

This is a book for anyone who develops software, including software engineers, scientists, engineers, and students of those disciplines. It is particularly suitable for scientists and engineers because of their frequent use of random numbers, either directly or indirectly, through the software they use on a daily basis. A poor choice of random number generator can prove catastrophic or at the least frustrating. Knowing something about how random number generators work will help avoid unfortunate consequences.

How To Use This Book

A basic reading of the book includes Chaps. 1, 2, and 4. These chapters introduce the concepts of pseudorandomness, the core set of uniform pseudorandom number generators, and how to test such generators.

Chapter 3 will be useful to those engaged in simulations of some kind. Read it after Chap. 2 as nonuniform pseudorandom number generation is based on uniform generators.

Chapter 5 covers methods for generating parallel streams of pseudorandom numbers. This chapter is best read after Chap. 2 and Chap. 4 on testing. While no examples are given specifically for graphics processors (GPUs), the methods are generic and apply to GPUs as well as distributed processes on CPUs.

Chapter 6 covering cryptographically secure pseudorandom number generators should also be read after Chaps. 2 and 4.

Chapter 7 on other random sequences can be read for fun, most profitably after Chaps. 2 and 4.

There are exercises at the end of each chapter. Most of these are small programming projects meant to increase familiarity with the material. Exercises that are (subjectively) more difficult will be marked with stars to indicate the level of difficulty.

Example code is in C and/or Python. For Python, we use version 2.7 though earlier 2.x versions should work just as well. Python 3.x will work with minor adjustments.

Intimate knowledge of these programming languages is not necessary in order to understand the concepts being discussed. If something is not easy to see in the code, it will be described in the text. Why C? Because C is a high-performance language that will make the example code all the more useful and because C is the grandfather of most of the common programming languages in current use, including Python.

Each code listing includes the name of the file containing the code. These files are available on the book website:

http://www.numbersandcomputers.com/random/

Note that the code listing in the book may be compressed to save space. The files on the website are as written and include comments and appropriate spacing.

For readers not familiar with C and/or Python, there are a plethora of tutorials on the web and reference books by the bookcase. Two examples, geared toward people less familiar with programming, are *Beginning C* by Ivor Horton and *Python Programming Fundamentals* by Kent Lee. Both of these texts are available from Springer in print or ebook format.

At the end of each chapter are references for the material presented in the chapter. Much can be learned by looking at these references. Almost by instinct we tend to ignore sections like these as we are now programmed to ignore advertisements on web pages. In this former case, resist temptation; in the latter case, keep calm and carry on.

Acknowledgments

In addition to my wife, Maria, I want to thank all of our children: David, Peter, Paul, Monica, Joseph, and Francis. Without your patience and encouragement none of this would have been written. Thank you for providing a living example of a random process. Life is truly more exciting the more it is shared.

Thornton, CO, USA
January 2018

Ronald T. Kneusel
AM+DG

Contents

Chapter 1
Random and Pseudorandom Sequences

Abstract Randomness is a fuzzy and difficult concept. In this chapter we side-step the philosophical issues and instead focus on random and pseudorandom sequences. We discuss what we mean by a random sequence and give examples of processes that generate randomness. We then conduct an experiment that shows humans are bad at randomness. Pseudorandom sequences are introduced next, along with an experiment showing that the quality of a pseudorandom sequence matters. We conclude with a quick look at hardware random number generation as supported by modern CPUs.

1.1 Random Sequences

What is randomness? A simple enough question to ask but answering it is difficult. It is somewhat similar to St. Augustine of Hippo's comment on time,

> What, then, is time? If no one ask of me, I know; if I wish to explain to him who asks, I know not. [1]

We all think we *know* what randomness is when we see it, but we cannot explain it.

Fortunately for us, we really do not need to know what randomness is, what we are really concerned with here is the idea of a *random sequence*. A thorough treatment of the concept is found in [2] but we will define a random sequence as,

> a sequence of numbers, n, within some bounded range, where it is not possible to predict n_{k+1} from any combination of preceding values n_i, $i = 0, 1, \ldots, k$.

We view a random sequence as a generator that gives a value when asked without us being able to predict that value based on any of the previous values.

Our definition has not specified a parent distribution. Random processes are often thought of as draws from a parent probability distribution so that even though we cannot predict n_k we can approximate the parent distribution by repeatedly drawing from the random process and building a histogram. Here "drawing" means asking the generator for another value. For our purposes, we assume unless otherwise stated

© Springer International Publishing AG, part of Springer Nature 2018
R. T. Kneusel, *Random Numbers and Computers*,
https://doi.org/10.1007/978-3-319-77697-2_1

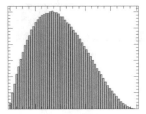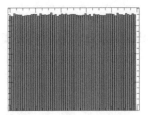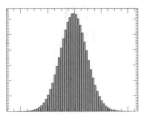

Fig. 1.1 Example distributions representing random processes with different parent distributions. Left, draws from a $Beta(2, 3)$ distribution. Center, a uniform distribution. Right, a Gaussian distribution with mean zero and standard deviation one ($N(0, 1)$). The shapes of the distributions are of interest here. The majority of the random number generators discussed in this book will assume a uniform parent distribution

that the parent distribution is uniform. This means that any value in the allowed range of values that could be returned by the generator is as likely as any other value.

Figure 1.1 shows histograms created by repeated sampling from a random process with a specific parent distribution. To be entirely transparent, the figures are based on the output of known good pseudorandom number generators, but we will for the moment take them as examples of actual random processes and the sequences they generate. Chapter 3 will talk more about random generators that do not have a uniform parent distribution. On the left is a histogram representing $Beta(2, 3)$. The center histogram is of a uniform distribution, while the histogram on the right represents a Gaussian with mean of zero and standard deviation of one ($N(0, 1)$).

Philosophy aside, we will assume that random sequences actually exist and are found in the physical world. For example, we take the following as examples of random processes from which one could generate a random sequence:

1. Flipping a fair coin.
2. Rolling a fair die.
3. The decay of radioactive elements.
4. The pattern of static on an old CRT television tuned to an empty channel.

In some sense, only number three above is truly random in that it is governed by quantum mechanics. Number four is also, to some extent. Still, for us mere mortals, any of these processes can be taken to be random and a potential source of a random sequence.

Most of us learned about probability in elementary school by using coins and dice. We will not review such basic things here other than to examine some coin toss sequences and ask questions about whether the results are what we would expect from a random process.

Below are two sequences generated by 26 flips a fair coin by two different subjects:

T T T T H T H T T T H H T H H T H T T T T H H H T H

and

H T T T H H H T T H T T T H H H T T H T H H T H H T

The coin in question was a pure copper United States penny from 1979.

The question is, are these sequences random? One way to answer the question is to use a statistical test to see if, given the null hypothesis that the number of tails should be 50%, we can to some precision reject the null hypothesis given the observed sequences. In other words, we want to calculate the probability of the observed deviation from 50% tails.

For the first sequence there are 11 heads and 15 tails. For 26 flips the null hypothesis says there should be 13 tails, therefore, the deviation is $15 - 13 = 2$ tails. The equation to calculate the probability is,

$$p = 2 \sum_{k=15}^{26} \binom{26}{k} \left(\frac{1}{2}\right)^{26} = 0.2786$$

because there were 26 trials and an "excess" of 2 tails over the expected 13. This is the binomial probability for an event with a 50-50 chance of happening summed over the sequence from 15 tails to 26 tosses. The factor of 2 is there because we need to account for the probability of the excess being a deficit.

From this calculation we get a p of 0.2786 which means that there is a better than 27% chance of seeing this sequence of heads and tails if the null hypothesis is true. In hypothesis testing, a p value of less than 0.05 is generally considered to be evidence that the null hypothesis is not true. In this case, we are well above the threshold so we accept the null hypothesis that this is a fair coin.

What about the other sequence? There were 13 heads and 13 tails. These are precisely the expected numbers so $p = 1.0$ and again we accept the null hypothesis and conclude that the coin is fair and that the likelihood of heads equals that of tails.

So, we have some evidence that, at least for the specific penny used above, we can treat the flipping of a coin as a random sequence. If we want to generate numbers we can use 0 for tails and 1 for heads and then group these binary digits into integers of any needed size. For example, if we needed values between 0 and 7 we could group the sequence in sets of 3 bits:

TTTTHTHTTTTHHTHHTHTTTTHHH =
$$\begin{aligned}
&= 000 \ 010 \ 100 \ 011 \ 011 \ 010 \ 000 \ 111\\
&= \quad 0 \quad \ 2 \quad \ 4 \quad \ 3 \quad \ 3 \quad \ 2 \quad \ 0 \quad \ 7
\end{aligned}$$

where have ignored the last two flips to have a sequence evenly divisible by three.

In truth, the flipping of a coin by a human is not entirely random as Diaconis et al., demonstrated in [3], but we will ignore this for the time being. Still, it is probably best not to bet the farm on the toss of a coin!

Number 4 in our list of random processes claims that we can use the noise pattern of an empty analog television channel to generate random numbers. Since much of the world has switched to digital television transmissions, including, finally, the United States, for the time being the old analog television frequencies are clear (for the most part). This means that the patterns seen on an old analog television screen tuned to an empty channel will be noise generated from whatever background signals may be present in the environment. This includes a small percentage of the cosmic background radiation itself [5].

After digging up an old Zenith analog television from the 1980s we took photographs of the screen when the TV was tuned to an empty station. For this experiment we used two different stations, one VHF and the other UHF. A sample photograph is seen in Fig. 1.2.

There were 149 such photographs taken with a digital camera. After cropping to 1500×1500 pixels to remove the television border and converting to floating point arrays using the NumPy library [6], the images were processed to generate a sequence of random bytes. At least, we hope they are random. The Python code to process the images is,

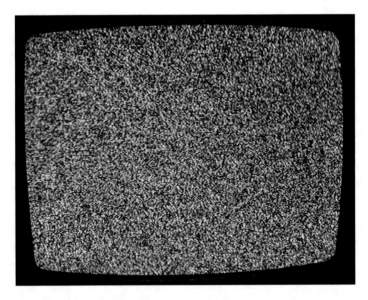

Fig. 1.2 A sample photograph of an old analog television in a dark room tuned to an empty station. A series of 149 such photographs were used to generate a sequence of random bytes. The television images were cropped to remove the screen border

```
1  import sys
2  import os
3  import numpy as np
4  from scipy.misc import imresize
5
6  def ProcessFrame(data):
7      m = data**2
8      m = (m - m.min())/(m.max() - m.min())
9      b = (255.0*m).astype("uint8")
10     b = imresize(b,(500,500),interp="nearest")
11     d = b.shape[0] if (b.shape[0] < b.shape[1])
                  else b.shape[1]
12     h = []
13     k = 0
14     for i in range(d):
15         for j in range(d):
16             h.append(int(b[i,j]) ^ int(b[d-j-1,i]) ^
                     int(b[j,d-i-1]) ^ int(b[d-j-1,d-i-1]))
17     v = []
18     vb = 0
19     c = 0
20     for n in h:
21         for j in range(4):
22             t = (n>>(2*j))&3
23             if (t==2):
24                 vb |= (1<<c)
25                 c += 1
26             elif (t==1):
27                 c += 1
28             if (c > 7):
29                 v.append(vb)
30                 vb = 0
31                 c = 0
32     return v
```
(process_frames.py)

The input to ProcessFrame is a 2D NumPy array representing a single band of the RGB digital image of the screen. Each band is processed separately. Lines 7 through 9 square the input and scale it to be in the range [0, 1] (line 8). Line 9 sets b to a byte array, [0, 255] and line 10 resizes it to 500 by 500 pixels.

Line 11 determines the smallest dimension of the input image. This allows us to work with non-square inputs. Lines 12 through 16 accumulate bytes in h. These are generated by XOR-ing portions of the image with itself. Lines 17 through 31 apply Von Neumann whitening. Von Neumann whitening operates on pairs of successive bits. If the two bits are the same, either 1-1 or 0-0, they are ignored. If the bits are 1-0 a 1 bit is output (line 23). If the bits are 0-1 a zero bit is output (line 26). When eight bits have been output the value is added to the output vector, v.

The `ProcessFrame` function is used by a small driver program that loads frames from a directory and applies the conversion to each image band while keeping all the returned byte arrays (lines 9–19),

```
1  def main():
2      fdir = "frames_tv/"
3      frames = [i for i in os.listdir(fdir) if
            i.find(".npy") != -1]
4      v = []
5      for f in frames:
6          d = np.load(fdir+f)
7          v += ProcessFrame(d[:,:,0])
8          v += ProcessFrame(np.flipud(np.fliplr(d[:,:,1])))
9          v += ProcessFrame(np.fliplr(np.flipud(d[:,:,2])))
10     with open("random_tv.dat","w") as f:
11         for i in xrange(len(v)):
12             f.write(chr(v[i]))
```
(process_frames.py)

where the final list of generated bytes (v) is written to a disk file (lines 10–12). Note, each band of the RGB image is flipped a different way before processing.

The code produces a file of bytes. How do we use them to see if they are randomly distributed? We will examine a plethora of techniques in Chap. 4 but one thing we can do here is simply apply them to a simulation and see if it gives us reasonable results. While this is not rigorous, it is perhaps more fun.

Our chosen simulation is the Monty Hall Dilemma. This dilemma is based on the old American television game show called Let's Make a Deal where the host, Monty Hall (really Monte Halparin), shows contestants three large closed doors on the stage numbered 1, 2, and 3. Behind one of the doors is a new car. Behind the other two are joke prizes like goats. Hall would ask the contestant to select a door. Say the contestant selected door number 1. He would then open one of the not selected doors, say door number 3, and ask the contestant if he or she wanted to change from selecting door number 1 to the other unopened door, door number 2. After making a final choice, Hall would open the contestant's selected door to reveal either a new car or a goat.

So, the question the simulation will address is whether the contestant should keep his or her original choice or change to the remaining door when given the opportunity? Will it matter or will it change nothing regarding the chance of winning the car?

In 1990 this dilemma was discussed in Marilyn Vos Savant's column in PARADE magazine. To say the least, it generated a great deal of controversy in which many mathematicians accused Vos Savant of providing an incorrect solution. A splendid account of this episode is found in [7] by Mlodinow as well as Vos Savant's own website [8].

Vos Savant's recommendation was for the contestant to change to the remaining door when given the opportunity. She claimed that doing so would allow the contestant to win the car $\frac{2}{3}$ of the time while, from basic probability, we see that simply selecting a door at random would win $\frac{1}{3}$ of the time on average.

We will use a small C program to simulate the game to see if Vos Savant was indeed correct. As a spoiler, she was, and after the simulation we will explain why she was correct.

The C program will read a stream of bytes from a file. It will use two bytes at a time to simulate a round of the game and after processing all the bytes in the file it will report the final tally. The program is,

```c
 1  int main(int argc, char *argv[]) {
 2      FILE *f;
 3      unsigned char b, win, guess;
 4      int i, n, niter, win0, win1;
 5
 6      win0 = win1 = 0;
 7      f = fopen(argv[1], "r");
 8      fseek(f, 0, SEEK_END);
 9      n = ftell(f);
10      fseek(f, 0, SEEK_SET);
11      niter = n / 2;
12
13      for (i=0; i < niter; i++) {
14          b = fgetc(f);
15          win = randint(b,3);
16          b = fgetc(f);
17          guess = randint(b,3);
18          if (win == guess) win0++;
19          switch (guess) {
20              case 0:
21                  if (win == 0) guess = 1;   // door 1 or 2
22                  if (win == 1) guess = 1;   // door 2
23                  if (win == 2) guess = 2;   // door 1
24                  break;
25              case 1:
26                  if (win == 0) guess = 0;   // door 2
27                  if (win == 1) guess = 0;   // door 0 or 2
28                  if (win == 2) guess = 2;   // door 0
29                  break;
30              case 2:
31                  if (win == 0) guess = 0;   // door 1
32                  if (win == 1) guess = 1;   // door 0
33                  if (win == 2) guess = 0;   // door 0 or 1
34                  break;
35          }
36          if (win == guess) win1++;
37      }
```

```
38        fclose(f);
39        printf("\nResults after %d simulations:\n\n", niter);
40        printf(" Wins on initial guess   = %d (%0.4f %%)\n",
              win0, 100.0*(float)win0/(float)niter);
41        printf(" Wins on changing guess = %d (%0.4f %%)\n",
              win1, 100.0*(float)win1/(float)niter);
42        printf("\n");
43        return 0;
44 }
```
(monty.c)

where lines 7 through 11 determine the length of the given file which determines the number of simulations (`niter`).

The `for` loop (lines 13–37) run the simulations by reading two bytes from the input file (lines 15 and 17) that are converted to an integer in the range [0, 2] by the `randint` function,

```
unsigned char randint(unsigned char b, unsigned char mx) {
    return (int)(mx*((float)b/256.0));
}
```

These represent the winning door number (`win`) and the contestant's initial guess (`guess`). Line 18 counts whether the initial guess would win the car or not.

We need to then examine three cases based on the contestant's initial guess. This is how we simulate Monty Hall and the selection of a door to open. One case will suffice for explanation. If the initial guess is door 0, we look to see what the winning guess is (lines 21–23). This is because Hall is not free to simply open just any door, he must open a door that is not the contestant's door nor the winning door, which he also knows. If the contestant picks the winning door initially (line 21) Hall has two choices and we simply select one of them. If the contestant picks a losing door, Hall has only one choice remaining (lines 22–23).

Let's pause here a moment. We have already spoiled things by stating that the contestant will win $\frac{2}{3}$ of the time by changing his or her initial guess. We can see a bit of why this is so with the case described above. Hall is not free to chose just any of the remaining doors unless the winning door and the contestant's initial door are the same (line 21). This means that the only remaining unopened door in the case when the winning door and the contestant's door are not the same must be the winning door. Consider line 22. The contestant has selected door 0, Hall has opened door 2 (as per the comment\), so if the contestant elects to change, the only remaining choice is door 1 (`guess = 1`).

Line 36 tracks whether changing doors results in a win. Lines 39–42 report on the final results of the simulations showing the fraction of times the initial guess was correct and the number of times changing to the remaining door is correct.

Let's run the program above on some streams of bytes. We have the stream of bytes from our television pictures. We also need a stream of truly random bytes with which to compare. We will ignore pseudorandom generated bytes for the moment but revisit this example a bit later once we've talked about pseudorandomness.

Where can we get a stream of truly random bytes? Our list of random processes above included the decay of radioactive material. This can be used to generate a stream of random bytes, for example, as are found on the HotBits website [9]. We downloaded 73728 such bytes to use them as input to the Monty Hall program. Let's run things and see where we end up.

If we start with the truly random bytes we get,

```
Results after 36864 simulations:

    Wins on initial guess  = 12389 (33.6073 %)
    Wins on changing guess = 24475 (66.3927 %)
```

whereas we get the following from our television images,

```
Results after 13793771 simulations:

    Wins on initial guess  = 4602527 (33.3667 %)
    Wins on changing guess = 9191244 (66.6333 %)
```

Clearly, we have only a limited amount of data in the true random case but we still see that changing our guess is converging towards the expected $\frac{2}{3}$ win rate. Encouragingly, our television generated data, of which we have a lot more, is also converging nicely towards the expected percentages. So, we have some *ad hoc* confidence that the sequence of bytes derived from images of an empty analog television signal are indeed useful as random numbers. We will, in Chap. 4, revisit these data sets and apply more rigorous randomness tests.

To finish this section let's see why Vos Savant was correct as our simulation clearly shows. We can all readily agree that simply choosing a door at random and not changing our choice will result in a win $\frac{1}{3}$ of the time. This is basic probability theory. The question, then, is why changing to the remaining door helps so much.

Our description of the simulation program above provides the key. Assume the contestant always changes to the remaining door, the one Hall has not opened. Now consider ways in which this choice results in a loss. If the initial selection was correct, changing from this means, naturally, that the contestant will lose. Hall will never open the door with the car. In this case Hall has two doors to choose from since neither has a new car behind it. However, $\frac{2}{3}$ of the time, the initial guess will be incorrect. What happens then? Since Hall is constrained in those cases to select only one possible door to avoid revealing the car, changing to the unopened door requires that the contestant will win, as the only unopened door must have a car behind it.

1.2   Experiment: Humans Are Bad at Randomness

Human beings evolved in a world where survival depended upon consistent, as opposed to random, action. We have eyes and brains that are quite sensitive to patterns and we expect things to just make sense and act consistently. This makes us bad at randomness. Humans are lousy random number generators as anyone who has tried to play the party game where each person thinks of a word entirely unrelated to any other words already spoken can attest. In this section we look at how humans do when asked to generate a random number in the range [1, 99].

The experiment was conducted using Amazon's Mechanical Turk service. Mechanical Turk lets people set up "HITs" (Human Intelligence Tasks) and have actual human beings ("Workers") perform the tasks for money. This service is often used by the academic community. In our case, a HIT was defined which simply asked the Worker to enter an integer in the range [1, 99]. For this task the worker was paid $0.02 USD. The experiment was actually run twice. The first run was restricted to workers in the United States while the second run was restricted to workers in India. In both cases, 500 subjects completed the experiment though some answers were not usable and were discarded.

It is natural to want to look at the distribution of numbers entered. These are shown in Fig. 1.3. Clearly, neither histogram is similar to a uniform distribution. This is not surprising, humans are bad at choosing random values. Problem 1.1 asks readers to conduct their own experiment along these lines. One question that can be asked is whether people are more likely to select an even or an odd number when asked for a random number in some range. As above for coins, the null hypothesis is that neither is more likely so we can use the same approach here that we used with coin flips.

For people from the United States, there were 500 total responses so we would expect 250 odd numbers if the null hypothesis is to be accepted. There were 273 odd numbers, therefore, the probability of seeing this deviation is,

$$p = 2 \sum_{k=273}^{500} \binom{500}{k} \left(\frac{1}{2}\right)^{500} = 0.0441$$

which is just below the typically accepted threshold of $p < 0.05$. So, we reject the null hypothesis and claim to have some evidence that people in the United States are more likely to select an odd number than an even number.

Let's repeat the calculation for the data from India. In this case we have 277 odd numbers for 476 usable responses. This gives us a p value of,

$$p = 2 \sum_{k=277}^{476} \binom{476}{k} \left(\frac{1}{2}\right)^{476} = 0.000406$$

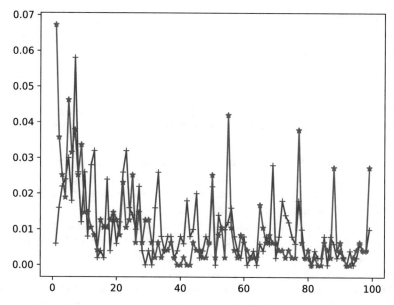

Fig. 1.3 Histograms showing the results of asking 500 Americans (blue, plus) and Indians (red, star) to enter a number between [1, 99]. Clearly, the values selected are not representative of a uniform distribution

which is highly significant showing that people from India are much more likely to select an odd number than an even number. According to the CIA World Factbook, 80% of the population of India identifies as Hindu. Hinduism has a rich numerology which emphasizes odd numbers. This may partly explain the results seen. Again, humans are bad at being random. This is good for us as a species but means we cannot rely on people to be a source of random number sequences.

1.3 Pseudorandom Sequences

We change now to considering pseudorandom sequences. Our definition above of a random sequence states that we can consider a sequence of numbers to be random if we cannot predict n_{k+1} from any combination of n_i, $i = 1, 2, \ldots, k$. Keeping this in mind we can now define a pseudorandom sequence:

A pseudorandom sequence is a deterministically generated sequence of numbers that is indistinguishable from a true random sequence of numbers.

Here the operative word is "indistinguishable". Chapter 4 will discuss in detail how to convince ourselves of this indistinguishability. For now we just assume that such a thing can exist, at least to the level necessary to be useful, and look at what happens when the pseudorandom sequence is closer or further from this ideal.

At all times in this book we must keep the words of John von Neumann in mind,

Any one who considers arithmetical methods of producing random digits is, of course, in a state of sin.

The quote is found in [4]. Most references of this quote neglect the remainder which is,

For, as has been pointed out several times, there is no such thing as a random number – there are only methods to produce random numbers, and a strict arithmetic procedure of course is not such a method.

This shows that our task is actually futile in that no sequence of deterministically generated numbers could ever be truly random. However, we press on because, while ultimately impossible, we can get asymptotically close to the ideal which is good enough for almost any purpose we have in mind.

We have a definition of a pseudorandom sequence but it should not feel satisfying because we have yet to specify *how* to generate such a sequence let alone how to test it. The answer to the *"how?"* question is Chap. 2 and the answer to the testing question is Chap. 4, but we can say some things here.

Firstly, we do not mean Fig. 1.4 when we mean pseudorandom number generator. Rather, let's focus on pseudorandom bytes. If we can generate those we are all set because we can group them to form random integers of any desired size. If we divide these random integers by one greater than their maximum value we can create floating-point values in the range $[0, 1)$. If we have floats in this range we can easily generate integers in any range we desire. If r is a random value $[0, 1)$ then a random integer in the range $[a, b)$, $a < b$ is,

$$i = \lfloor r(b - a) \rfloor + a$$

One of the first attempts to generate a pseudorandom sequence of numbers was the "Middle Square" method of Von Neumann [11]. The idea is simple. If we want to generate 32-bit integers we start with a 32-bit integer, square it, store it in a 64-bit integer, and return the middle 32-bits. Code to generate a file of such values is,

```
int getRandomNumber()
{
    return 4; // chosen by fair dice roll.
              // guaranteed to be random.
}
```

Fig. 1.4 How not to generate pseudorandom numbers. From http://xkcd.com. Used under the Creative Commons Attribution-NonCommercial 2.5 License along with the additional use cases defined here: https://xkcd.com/license.html

```
1  #include <stdio.h>
2  unsigned long long seed = 0xfedcb2ed;
3
4  unsigned long random(void) {
5      seed *= seed;
6      seed = (seed >> 16) & 0xffffffff;
7      return (unsigned long)seed;
8  }
9
10 int main(int argc, char *argv[]) {
11     int i;
12     FILE *f;
13     unsigned long n;
14
15     f = fopen(argv[2],"w");
16     for(i=0; i<atoi(argv[1]); i++) {
17         n = random();
18         fwrite((void *)&n, 4, 1, f);
19     }
20     fclose(f);
21     return 1;
22 }
```
(middle.c)

where the initial seed value, the start of the sequence, is in seed. The function random is where the action is. We want 32-bit values so we square the seed, storing it in a 64-bit integer (line 5) and update it with the middle 32-bits (line 6). We return this as our random integer and store it on disk.

If we run this program to create two million 32-bit values and use the output with our Monty Hall program we get,

```
Results after 4000000 simulations:

  Wins on initial guess  = 3990003 (99.7501 %)
  Wins on changing guess = 9997 (0.2499 %)
```

which is clearly wrong. If we trusted this generator, we'd never change our answer. The middle square method is seemingly a lousy pseudorandom generator, and it is. But, it can be regarded as a first attempt of sorts by the computer science community. Still, it illustrates the essential idea of a pseudorandom generator: apply some process, based on some seed, that results in a sequence of values hopefully approximating a truly random sequence. One reason why the middle square method is so poor is that a middle sequence of zero will be zero forever. Examining the bytes in the output file shows that this is indeed the case part way in.

Interestingly, just recently (as of 2017), the middle square method has resurfaced. In [12] Widynski proposes a straightforward modification of the middle square algorithm which results in a very fast generator with a respectable period, excellent results on randomness tests, and easy application in parallel. We show here a slightly modified version of the example program to expand some of the compactness for pedagogical reasons,

```
 1  #include <stdio.h>
 2  #include <stdint.h>
 3  uint64_t seed = 0xb5ad4eceda1ce2a9;
 4
 5  uint32_t random(void) {
 6      static uint64_t x=0, w=0;
 7      x *= x;
 8      w += seed;
 9      x += w;
10      x = (x >> 32) | (x << 32);
11      return (uint32_t)x;
12  }
13
14  int main(int argc, char *argv[]) {
15      int i;
16      FILE *f;
17      uint32_t n;
18
19      f = fopen(argv[2],"w");
20      for(i=0; i<atoi(argv[1]); i++) {
21          n = random();
22          fwrite((void *)&n, 4, 1, f);
23      }
24      fclose(f);
25      return 1;
26  }
```
    (middle_weyl.c)

where we have also included stdint.h so that we can be very explicit about the size of each integer value.

Line 3 sets the seed value. It must be odd and the upper 32-bits cannot be zero. The value here is from the example in [12]. We make it global so that it can be updated, if desired. Lines 14–26 are a small driver program to generate a desired number of 32-bit integers and store them on disk.

The function random does all the work. N.B. that x and w are static variables, they are initialized once to zero and then preserved between calls to random. A more robust version would make these values accessible so the sequence could be reset and a different seed used. Line 7 squares x while line 8 adds the seed to the running sum in w. Line 9 adds w to x while line 10 swaps the middle part of the square of x so that it is in the lower 32-bits. Line 11 returns x where the cast preserves only the lower 32-bits. This will return the middle part of x and is why this generator is still a middle square generator. The value w implements a Weyl sequence that avoids the middle zero problem we saw above. Because of this, we will call this the "Middle Weyl" generator (see Chap. 4) though it is called "msws" in [12].

Compiling, running, and passing the output to our Monty Hall program produces,

```
Results after 4000000 simulations:

  Wins on initial guess  = 1333016 (33.3254 %)
  Wins on changing guess = 2666984 (66.6746 %)
```

which is as good a result as any we have seen in this chapter. We will revisit this generator in Chap. 4 and subject it to additional tests.

1.4 Experiment: Fractals and Good Versus Bad Pseudorandom Values

At the heart of this book is the notion that pseudorandom number generators can be good or bad in relation to some purpose for which we desire a sequence of random numbers. In this section we present a fun and attractive use for random numbers: the generation of fractal images via the Iterated Function System (IFS) of Barnsley et al. [13, 14].

A fractal is a mathematical object which exhibits self-similarity, i.e., view the fractal at one scale and again at another scale and the images look similar. One way to generate fractal images is to define a series of contractive mappings of the unit square ([0, 1),[0, 1)) onto itself. This is what IFS does. In particular, we will experiment with the 2D version of the Random Iteration Algorithm from [14]. It is beyond our purposes here to delve into what, exactly, is happening to generate the fractals, but the essence of the algorithm is,

1. Pick an starting point in $\mathbb{R}^2$, p_0, within the unit square and form a vector: $[p_x, p_y, 1]$.
2. From a collection of matrices defining a particular fractal, select one, M_i, with probability P_i where $\sum_i^N P_i = 1$. This is a 3×3 matrix.
3. Calculate $p_1 = M_i\, p_0$ and plot the x and y components of p_1.
4. Repeat from Step 1 selecting maps at random and using the previous point p_k to calculate p_{k+1}.

The first N points are generally thrown away to allow the iterated system time to approach the fractal attractor. In practice this means annoying points that don't "fit" the pattern will be discarded.

We can implement this algorithm simply in Python using NumPy by defining an IFS class. The definition of the class and the fractals comes first,

```
1  import sys
2  import numpy as np
3  class IFS:
4      maps = {
5          "circle": {"nmaps":4, "probs": [0.25, 0.25,
                  0.25, 0.25],
6                       "maps": [[[.2929, -.2929, .5],
                               [.2929, .2929, .5], [0,0,1]],
7                               [[-.2929, .2929, .5],
                               [-.2929, -.2929, .5], [0,0,1]],
8                               [[.2929, .2929, .5],
                               [-.2929, .2929, .5], [0,0,1]],
9                               [[-.2929,-.2929,.5],
                               [.2929,-.2929,.5], [0,0,1]]]},
10         "dragon": {"nmaps":2, "probs": [0.5, 0.5],
11                       "maps": [[[.5, -.5, 0], [.5, .5, 0],
                               [0,0,1]],
12                               [[-.5, -.5, 1], [.5,-.5,0],
                               [0,0,1]]]},
13         "fern":   {"nmaps":4, "probs":[0.75, 0.115,
                  0.115, 0.02],
14                       "maps":[[[.81, .07, .12],
                               [-.04, .84, .195], [0,0,1]],
15                               [[.18, -.25, .12],
                               [.27, .23, .02], [0,0,1]],
16                               [[.19, .275, .16],
                               [.238, -.14, .12], [0,0,1]],
17                               [[.0235, .087, .11],
                               [.045, .1666, 0], [0,0,1]]]},
18         "koch":    {"nmaps":4, "probs":[0.25,0.25,0.25,0.25],
19                       "maps":[[[0.333, 0, 0], [0, 0.333, 0],
                               [0,0,1]],
20                               [[0.167, -.288, .333],
                               [.288, .167, 0], [0,0,1]],
21                               [[.167, .288, .5],
                               [-.288, .167, .288], [0,0,1]],
22                               [[.333, 0, .667],
                               [0, .333, 0], [0,0,1]]]},
23         "shell": {"nmaps":2, "probs":[0.955, 0.045],
24                       "maps":[[[.96, .15,-.08],
                               [-.15, .96, .0861], [0,0,1]],
25                               [[.11, -.05, 1.11],
                               [.05, .11, .8], [0,0,1]]]},
26         "sierpinski": {"nmaps":3, "probs":
                  [0.3333,0.3333,0.3333],
27                       "maps":[[[.5, 0, 0], [0, .5, 0], [0,0,1]],
28                               [[.5, 0, .5], [0, .5, 0],
                               [0,0,1]],
```

```
29                          [[.5, 0, .25], [0, .5, .5],
                                [0,0,1]]},
30          "tree": {"nmaps":6, "probs":
                    [.05, .05, .4, .05, .05, .4],
31              "maps":[[[.03, 0, -5e-3], [0, .48, 0],
                            [0,0,1]],
32                      [[.035, 0, .01], [0, -.44, .21],
                            [0,0,1]],
33                      [[.41, -.41, -0.1], [.41, .41, .23],
                            [0,0,1]],
34                      [[.03, 0, .01], [0, .48, 0],
                            [0,0,1]],
35                      [[.035,0, -7e-3], [0, .41, .02],
                            [0,0,1]],
36                      [[.41, .41, .01], [-.41, .41, .23],
                            [0,0,1]]]}
37      }
(ifs.py)
```

where the maps are 3×3 matrices ("maps") which will be selected by the given probabilities ("probs"). Each map updates the current point $p_k = [x, y, 1]$. For example, one of the fern maps updates p_k via,

$$p_{k+1} = \begin{bmatrix} 0.81 & 0.07 & 0.12 \\ -0.04 & 0.84 & 0.195 \\ 0 & 0 & 1 \end{bmatrix} \begin{bmatrix} x \\ y \\ 1 \end{bmatrix}$$

The methods of the IFS class come next,

```
1  def Random(self):
2      if (self.gen == "good"):
3          self.x = 16807*self.x % 2147483647
4          m = 2147483647
5      else:
6          self.x = 259*self.x % 32768
7          m = 32768
8      return self.x/float(m+1)
9
10 def ChooseMap(self):
11     r = self.Random()
12     a = 0.0
13     k = 0
14
15     for i in range(self.nmaps):
16         if (r > a):
17             k = i
```

```
18          a += self.probs[i]
19      return k
20
21 def GeneratePoints(self):
22      self.xy = np.zeros((self.npoints,3))
23
24      xy = np.array([self.Random(), self.Random(), 1.0])
25      for i in xrange(100):
26          m = self.maps[self.ChooseMap(),:,:]
27          xy = np.dot(m,xy)
28      for i in xrange(self.npoints):
29          k = self.ChooseMap()
30          m = self.maps[k,:,:]
31          xy = np.dot(m,xy)
32          self.xy[i,:] = [xy[0],xy[1],k]
33
34 def StorePoints(self):
35      np.savetxt(self.outfile, self.xy)
36
37 def GetPoints(self):
38      return self.xy
39
40 def __init__(self, npoints, outfile, name, gen):
41      self.gen = gen
42      self.x = 1  # always start at 1
43      self.npoints = npoints
44      self.outfile = outfile
45      self.name = name
46      self.nmaps = self.maps[name]["nmaps"]
47      self.probs = np.array(self.maps[name]["probs"])
48      self.maps = np.array(self.maps[name]["maps"])
```
(ifs.py)

The class generates the fractal points in 2D and stores them. The methods
GetPoints and StorePoints return or write the points to disk, respectively. The
constructor (lines 40–48) simply stores input parameters and convenience copies of
the maps for the desired fractal. Of interest is line 42 that sets the pseudorandom
number seed to 1. We will see in Chap. 2 precisely what this means but for now it
means that the sequence of numbers generated will always be the same. Line 43 sets
the number of points to generate. A good value is at least 100,000 or more.

Lines 21–32 define the heart of the IFS class. Line 22 defines the output array.
There are npoints points each of which is an *x* and *y* location. The last column of
the array stores the map number which defined the point. As we will see, if we track
the map which defines the point we can generate a color representation of the fractal
showing each map clearly.

   Line 24 creates our initial point by calling the Random method to return a floating-point number, [0, 1). This method will use one of two pseudorandom number generators, a "good" one or a "bad" one as selected by the user. Lines 25–27 and 28–32 are identical in that they iterate selecting maps at random according to their probability of being selected (ChooseMap) and multiplying the current point by that map. The only difference is that the first for loop discards the first 100 points while the second preserves the next npoints points. This removes the transient part of the fractal. The NumPy library uses np.dot for matrix multiplication (lines 27 and 31). So, when GeneratePoints is finished, the member variable xy will hold all the points of the fractal along with the map number that generated that point.

   The ChooseMap method selects one of the fractal maps according to its probability of being selected. For example, the Sierpinski triangle defines three maps and associates a probability of $\frac{1}{3}$ to each of them while the shell fractal defines two maps with very unbalanced probabilities of 0.955 and 0.045 so 95% of the time, for the shell fractal, ChooseMap will return the first map.

   Lines 1–8 defining Random are of primary interest to us. This method defines two linear congruential generators (LCG) which are described in detail in Chap. 2. LCG generators are simple to implement but depend critically on their parameters in order to perform well. The "good" generator uses the parameters found by Park and Miller [15] which are adequate for our purposes. The "bad" generator uses parameters for a 16-bit generator with a period that will not be adequate for generating IFS fractals. The period of a generator is the number of values it will generate before repeating. The Park and Miller generator has a period of about 2^{31} which is well below the number of values we will be using. By way of contrast, the other generator has a period of just above 8200 (found empirically).

   A short driver program will complete our IFS fractal application,

```
 1 def main():
 2     gen = sys.argv[1].lower()
 3     npoints = int(sys.argv[2])
 4     outfile = sys.argv[3]
 5     name = sys.argv[4]
 6
 7     if (name not in ["sierpinski","fern","tree","circle",
                 "shell","dragon","koch"]):
 8         print "Unknown fractal: %s" % name
 9         return
10
11     app = IFS(npoints, outfile, name, gen)
12     app.GeneratePoints()
13     app.StorePoints()
```
(ifs.py)

where a sample call to create the fern fractal might be,

```
$ python ifs.py good 100000 fern.txt fern
```

Fig. 1.5 Fractals generated via the iterated function system (IFS). The colors represent the different contractive mappings used

which will use the Park and Miller generator to create `fern.txt`, a text file of 100,000 lines defining the fern fractal. The first line of this file is,

```
5.434129706450262498e-01 7.415229660632343123e-01 0.000e+00
```

showing that the first point of the fractal is approximately (0.5434, 0.7415) and map 0 was used.

The collection of possible fractals, each generated with 1,000,000 points and the Park and Miller generator, is shown in Fig. 1.5. The colors represent the different maps that make up the fractal.

The Python program below will generate the plots as seen in Fig. 1.5. It uses the Matplotlib library which is not part of the standard Python library but, like NumPy, is common and generally easily installed. The code is,

```python
 1  import numpy as np
 2  import matplotlib.pylab as plt
 3  import sys
 4
 5  def main():
 6      points = np.loadtxt(sys.argv[1])
 7
 8      if (len(sys.argv) < 3):
 9          output = ""
10      else:
11          output = sys.argv[2]
12
13      colors = ['g','c','m','y','b','k','0.5']
```

```
14      clr = []
15      for i in range(points.shape[0]):
16          clr.append(colors[int(points[i,2])])
17
18      plt.axis("equal")
19      plt.scatter(points[:,0], points[:,1], c=clr,
            marker=".", s=1)
20
21      if (output != ""):
22          plt.savefig(output, dpi=1000)
23
24      plt.show()
25
26  main()
```
(ifs_plot.py)

where the program is called with up to two arguments. The first, required, is the name of the output points file generated by the IFS program above. The second, if present, is the name of an output graphics file. So, to plot the points of the fern generated above use,

```
$ python ifs_plot.py fern.txt fern.png
```

which will show an interactive plot and also save the plot as a PNG file.

Now that we know how to generate fractals, let's look at the effect of the pseudorandom generator on the output. We will use the fern example and look at the generated fractal at two levels. The first is the entire fractal and then at a zoomed in region. In both cases the fractal will plot 1,000,000 points, the only difference will be the generator used: the "good" Park and Miller or the "bad" 16-bit generator. If we do this, we get the plots in Fig. 1.6 where it is plain to see that the good generator has filled in the space well enough to see zoomed in detail but the bad generator, which has a short period and only gives a small number of values, has failed to do anything beyond hint at the richness of the fractal.

The period of a generator is an important property but, as we will see in Chap. 4, there are other things to consider before deciding that a pseudorandom generator is fit for a particular purpose.

1.5   A CPU Hardware Generator

This chapter would be incomplete if we did not take at least a cursory look at the sort of random numbers modern CPUs are now capable of generating in hardware. Here we look at the RDRAND instruction introduced into Intel and AMD CPUs in the 2014–2015 time-frame. This instruction uses a hardware pipeline with an entropy source seeding a strong cryptographically secure pseudorandom number

Fig. 1.6 The difference a good pseudorandom generator makes is shown here by comparing the fern fractal generated with the Park and Miller generator (top) and a poor 16-bit LCG generator (bottom). In each case one million points were generated but because of the very low period of the bad generator, approximately 8200 values, only a shadow of the fractal is visible and when zoomed in (right) only a few points are visible revealing nothing of the rich fractal structure

generator to calculate random numbers on the fly. In a sense, even though a hardware instruction, RDRAND is actually a hybrid that is similar to some of the combined generators we will see later in this book.

Under some versions of the Linux operating system, the RDRAND instruction is used by the /dev/random and /dev/urandom pseudo files for random number generation. We will ignore these ways of accessing random numbers here and focus instead on direct use of RDRAND via C. We will also ignore the (not so insignificant) concerns that hardware instructions like RDRAND may be compromised by state-level agencies in order to weaken applications based on cryptography.

Our inspiration comes from the Intel Digital Random Number Generator Software Implementation Guide [16]. This guide details the operation of the hardware generator and offers guidance on how to add the functionality to your own programs. We will use this information to write a simple C program to generate a file of random bytes and then pass them to the Monty Hall program above to see how well they perform. The output of this program will also be suitable for those completing Problem 1.2. The code is straightforward,

```
 1 #include <stdio.h>
 2
 3 int random(unsigned long long *rnd) {
 4     int i;
 5
 6     for (i=0; i<10; i++)
 7         if (__builtin_ia32_rdrand64_step(rnd))
 8             return 1;
 9     return 0;
10 }
11
12 int main(int argc, char *argv[]) {
13     int i;
14     FILE *f;
15     unsigned long long r;
16
17     f = fopen(argv[2],"w");
18     for(i=0; i<atoi(argv[1]); i++) {
19         if (random(&r)) {
20             fwrite((void *)&r, 8, 1, f);
21         } else {
22             return 0;
23         }
24     }
25     fclose(f);
26     return 1;
27 }
```
(drng.c)

and must be compiled on a 64-bit system which supports the RDRAND instruction like so,

```
$ gcc drng.c -o drng -mrdrnd
```

We can then create a file of random bytes,

```
$ drng 2000000 random_rdrand.dat
```

and see what our Monty Hall program makes of them,

```
Results after 8000000 simulations:

   Wins on initial guess  = 2667432 (33.3429 %)
   Wins on changing guess = 5332568 (66.6571 %)
```

showing that, at least for this particular example, we do get good performance as expected.

The program itself opens the given output file name (line 17) and loops requesting the specified number of 64-bit random integers (lines 18–24). The function `random` updates the given 64-bit integer with a random value which, if successful, is written to the output file (line 20).

The `random` function is a bit funny. Why is there a `for` loop when all we want is a single random value? The RDRAND instruction uses a hardware generator that is updating the seed of the hardware-implemented pseudorandom generator based on an entropy source in the CPU. So, every time the `__builtin_ia32_rdrand64_step` function, which uses RDRAND, is called, there may not actually be a value ready to read. The loop tries ten times which is recommended since, according to the binomial distribution, the likelihood of a value never being ready is astronomically small in that case. If it does happen, the function will return zero and the program will halt.

1.6   Chapter Summary

In this chapter we introduced what we mean by random and pseudorandom sequences. We gave examples of true sources of randomness and ran an experiment using one of them in order to generate a sequence of random bytes. We then used a simple simulation (Monty Hall Dilemma) to see that the sequence of bytes performed reasonably. We showed with another experiment that humans are poor at generating random numbers. We demonstrated the effect of a poor pseudorandom sequence on a simulation (fractals) and took a quick look at modern CPU generation of random numbers.

Exercises

1.1 One of the experiments in this chapter asked people to randomly choose a number in the range [1, 99]. Another approach would be to ask people to generate a sequence of random digits. Ask at least 10 people to generate a sequence of 100 random digits in the range [0, 9]. Then, use the number of odd digits and the analysis techniques of this chapter to calculate a p-value as to whether or not it is reasonable to believe the sequences are random. You will get a p-value for each subject and a final p-value by pooling all the data from all subjects.

1.2 Modify the IFS fractal generating program above to read four bytes at a time from a file, convert them into an unsigned 32-bit integer and divide by 2^{32} to make a floating-point number in the range [0, 1). Alter the Random method to use these numbers and generate fractals with the television image data which is available on the book website (http://www.numbersandcomputers.com/random/). *

1.3 The LavaRnd website [10] details how to use the thermal noise of a standard webcam to generate a sequence of random numbers. Build a LavaCan™ according to the instructions on the LavaRnd website and use it to generate a file of random bytes. Run these against the Monty Hall program to convince yourself that the sequence is random. ***

References

1. Augustine of Hippo, Confessiones lib xi, cap xiv, sec 17, circa 400 AD.
2. Volchan, Sergio B. "What is a random sequence?." The American mathematical monthly 109.1 (2002): 46–63.
3. Diaconis, Persi, Susan Holmes, and Richard Montgomery. "Dynamical bias in the coin toss." SIAM review 49.2 (2007): 211–235.
4. Forsythe, G. E., H. H. Germand, and A. S. Householder. "Monte carlo method." NBS Applied Mathematics Series 12 (1951).
5. Cheng, Ta-Pei, and Brian H. Benedict. A college course on relativity and cosmology. Oxford University Press, 2015.
6. http://www.numpy.org/.
7. Mlodinow, Leonard. The drunkard's walk: How randomness rules our lives. Vintage, 2009.
8. http://marilynvossavant.com/game-show-problem/.
9. http://www.fourmilab.ch/hotbits/.
10. http://www.lavarand.org/.
11. John von Neumann, "Various techniques used in connection with random digits," in A.S. Householder, G.E. Forsythe, and H.H. Germond, eds., Monte Carlo Method, National Bureau of Standards Applied Mathematics Series, vol. 12 (Washington, D.C.: U.S. Government Printing Office, 1951): pp. 36–38.
12. Bernard Widynski, "Middle Square Weyl Sequence RNG", arxiv.org, https://arxiv.org/abs/1704.00358.
13. Barnsley, Michael F., and Stephen Demko. "Iterated function systems and the global construction of fractals." Proceedings of the Royal Society of London A: Mathematical, Physical and Engineering Sciences. Vol. 399. No. 1817. The Royal Society, 1985.
14. Barnsley, Michael F. Fractals everywhere. Academic press, 2014.
15. Park, Stephen K., and Keith W. Miller. "Random number generators: good ones are hard to find." Communications of the ACM 31.10 (1988): 1192–1201.
16. https://software.intel.com/en-us/articles/intel-digital-random-number-generator-drng-software-implementation-guide.

Chapter 2
Generating Uniform Random Numbers

Abstract Integers uniformly distributed over some interval are at the heart of pseudorandom number generators. This chapter examines techniques for generating pseudorandom integers: some historical, some useful for noncritical tasks such as games, and others which are workhorses and should be thought of as go-to methods. We will also take a look at how combining generators can increase their periods. We end the chapter with a brief look at quasirandom generators.

2.1 Uniform Random Numbers

The generation of uniformly distributed pseudorandom integers over some range is the core operation of most pseudorandom number generators. In this chapter we will examine several uniform random number generators. Some are important because they are historic while others are in common use and still others deserve to be more widely known. All of them produce uniformly distributed integers over some range. We will also look at combining generators to increase their period and look briefly at quasirandom processes.

To be explicit, we define a uniform pseudorandom number generator, $\mathbb{P}$, as a process which produces a sequence of integers, n, over some range, $[0, m)$, $0 \leq n < m$, for some maximum value m, never achieved, such that our definition of a random sequence from Chap. 1 holds,

> A random sequence is a sequence of numbers n, within some bounded range, where it is not possible to predict n_{k+1} from any combination of preceding values n_i, $i = 0, 1, \ldots, k$.

We approach this definition asymptotically by which we mean $\mathbb{P}$ passes randomness tests (see Chap. 4) to the level necessary for our purposes.

We write $[0, m)$ as the range so that we can easily form floating-point values from our uniform generator,

$$f = \frac{\mathbb{P}}{m}, \; f \in [0, 1)$$

which through simple scaling gives us floating-point numbers in any range,

$$w = a + (b - a) * f, \; w \in [a, b)$$

We will make use of these equations many times and indeed have already done so in Chap. 1 (see Sect. 1.1, Monty Hall Dilemma).

Since there are far too many generators to investigate in detail, indeed, a "bestiary" of pseudorandom number generators would be perhaps an interesting project, we will discuss only a few key generators in detail. These include linear congruential generators and the Mersenne Twister. The former is of historical interest and is still useful for certain applications such as games while the latter is the most commonly used generator at present. We should note that we are not ignoring the very important subject of cryptographically secure pseudorandom number generators (CSPRNGs) but are instead deferring them to Chap. 6 so as to not distract from the goal of this chapter. The other generators we investigate are xorshift and its variants, and the complementary multiply-with-carry generator with its exceptionally long period. We round out the chapter by looking at combining generators and take a quick look at quasirandom generators which are space filling at the expense of passing statistical randomness tests.

Note that in our definition above we did not specify the size of m. If we have a generator that can create bytes, then $m = 2^8 = 256$ while a generator that creates unsigned 32-bit integers has $m = 2^{32} = 4294967296$. If we want fine control over the number of floating-point values we can create from our generator we naturally want m to be as large as is possible up to the limitations of the IEEE 754 floating-point format for single or double precision floats. If we want single precision, then we probably want $m = 2^{32}$ so that we can take full advantage of the 24 bit significand. Likewise, for double precision, we probably want $m = 2^{64}$ to cover the 53 bit significand.

For example, in Sect. 1.1 we used analog television images to create a file of randomly distributed bytes. We then used these bytes, one at a time, to simulate the Monty Hall dilemma and showed that we achieved good results. We did so because we needed to select from only three doors so dividing by 256 covered our range very finely, so finely that when we cast back to an integer we were able to uniformly cover the 0–2 range we needed. However, let's now use this file of random bytes so see how m affects our ability to use floating-point random numbers.

We can use random floating-point numbers to estimate the value of π. How? By comparing areas, specifically, by comparing the area of a square of length one to one quarter of a circle of radius one. A square of length one starting at the origin has an area of one since $1 \times 1 = 1$. The area of a circle of radius one centered at the origin has an area of $\pi(1)^2 = \pi$ which means that the area of the circle in the first quadrant is $\pi/4$. Dividing the area of the circle in the first quadrant by the area of the square gives $(\pi/4)/1 = \pi/4$. Therefore, if we select pairs of random floating-point numbers [0, 1) and count the number that fall within the circle, we can estimate the area of the circle in the first quadrant. Since all those points immediately fall within the square we know that the ratio of the number of randomly selected points falling within the circle divided by the number of randomly selected points will approach $\pi/4$ assuming the points are uniformly distributed over the unit interval. We will again use the file of bytes we generated in Chap. 1 from the old analog television images. What we are testing here is the effect of the number of bytes that we use to generate a floating-point number on the distribution of the floating-point values we can generate (i.e. the effect of the size of m above).

First we need a simple program to estimate π using one, two, three or four bytes at a time from the file,

```
1   #include <stdio.h>
2   #include <stdint.h>
3   #include <math.h>
4
5   int in_circle(FILE *f, int s) {
6        uint64_t i, a=0, b=0;
7        double x,y;
8
9        for (i=0; i<s; i++) {
10            a <<= 8;
11            a |= (int)fgetc(f);
12            b <<= 8;
13            b |= (int)fgetc(f);
14       }
15
16       x = (double)a / (1ULL << (8*s));
17       y = (double)b / (1ULL << (8*s));
18       return (sqrt(x*x+y*y) <= 1.0);
19  }
20
21  void main(int argc, char *argv[]) {
22       FILE *f;
23       int i,s,c=0,k=1000000;
24
25       s = atoi(argv[1]);
26       f = fopen(argv[2], "r");
27
28       for (i=0; i < k; i++)
29            c += in_circle(f,s);
30
31       fclose(f);
32       printf("Estimated value of pi: %0.7f\n\n",
                   4.0*((double)c/k));
33  }
```
(random_pi.c)

This program uses the command line to specify first how many bytes at a time to use followed by a file of random bytes. Note, we do not check if there are actually enough bytes in the file to generate 1,000,000 pairs of floating-point numbers so, if using a size of 4, ensure that the file has at least 8 million bytes in it.

Lines 21 through 26 are straightforward, simply get the number of bytes to use (s) and open the file (line 26). Note that we set the iteration limit in line 23 (k=1000000). Lines 28 and 29 iterate by calling in_circle which returns 1 if the next point from the file falls within the circle or 0 if not. There is no need to count the points that fall within the square as we know they all will. Line 32, then, reports the outcome. There are c points within the circle and k points within the square. Since we know that ratio is $\pi/4$i, we multiply by 4 to get an estimate of π.

The function in_circle appears a little cryptic at first. We pass in the file pointer (f) and the number of bytes to use at a time (s). We use unsigned 64-bit integers here, initialized to zero (a, b), to create an integer of s bytes read from the input file. These define x and y which in turn define a point in the first quadrant. Lines 9 through 14 create the proper sized integers. Each new byte read is OR-ed into either a or b after shifting up by 8 bits to make room for it. Lines 16 and 17 turn these integers into floating-point numbers by dividing by one larger than the largest value the integer could take. This makes x and y fall into the range [0, 1). The expression 1ULL << (8*s) shifts 1 up by the number of bits corresponding to the number of bytes we are using.

Finally, line 18 simply asks if the point (x, y) falls within a circle of radius one meaning is the distance from the origin to the point (x, y) less than or equal to one. It is the truth value of this expression, either 1 or 0, that is added to our count in line 29.

If we compile this program, recalling that we need to add -lm if using gcc, we get the following output for 1, 2, 3, or 4 bytes at a time, which is reasonable when

Bytes	Estimate of π
1	3.1577360
2	3.1428880
3	3.1424840
4	3.1443440

using two or more bytes at a time as each is less than 0.1% off from the true value (to eight places). However, using only one byte at a time, we jump to an error of 0.5% indicating that 8 bits is not sufficient in this case. Why? Because since we are taking pairs of numbers to select a point we are therefore only capable of selecting at most $256^2 = 65536$ different points distributed over the unit interval ([0, 1), [0, 1)). If we move to four bytes at a time we increase this to $(2^{32})^2$ possible points should we have enough random data. Naturally, this will lead to a vastly improved estimate of π. So, even at four bytes at a time, why isn't our estimate all that good? One factor is using only 1,000,000 iterations. If we had the data (we don't in this case) and could use ten times as many iterations we should improve our result. Another, likely more important, factor is that our file of bytes isn't entirely random. We will test this more in Chap. 4 but we can see it here as a plausible statement if we replace the TV file with a series of bytes generated by a known generator with very good performance, in this case the generator in the OpenSSL library. If we do this, for the same number of iterations and four bytes at a time, we get $\pi \approx 3.1414520$ for an error of 0.004%. Indeed, the ent Unix program for quickly testing the randomness of a file uses an estimate of π as one of its reported outputs.

So, we now know how to work with floating-point values generated from sequences of random integers. We also know to be a bit careful when converting between random integers and floating-point values to ensure that we have enough bits to create enough floats to accurately model what it is we want to model. Let's move ahead and look at our first pseudorandom number generator.

2.2 Linear Congruential Generators

A mixed linear congruential generator (mixed LCG) is an algebraic method using modular arithmetic that produces sequences of numbers which, when carefully tuned, can pass enough randomness tests to be useful. In particular, the generator uses a recurrence relation based on a seed value (x_0) so that the next number in the sequence is,

$$x_{n+1} = (ax_n + c) \bmod m$$

where a, $0 < a < m$ is the multiplier, c, $0 \le c < m$ is the increment and m is the modulus. It is important to note that a, c and m are all integers and only integer arithmetic is used in the calculation. If we want floating-point, we simply scale by m, this time using floating-point arithmetic: $f_n = x_n/m$.

By careful selection of a, c and m one is able to produce a usable sequence of integers. We will see below how choosing these values affects the sequence generated. We can already see an upper limit on the period, or length, of any sequence generated. It is as most m, since we are modulo m, but it might well be considerably less than m.

If $c = 0$ we have a multiplicative LCG. Multiplicative linear congruential generators were introduced in 1949 by Lehmer [1]. This is described in detail along with other interesting historical notes in the more readily accessible 1962 paper by Hull and Dobell [2]. The Park and Miller generator used in Chap. 1 is really a Lehmer-style multiplicative generator.

Let's look at how a simple LCG behaves. The program below generates sequences of numbers for an LCG with $m = 10$ and all allowed values of a, c, and s the initial seed. It also calculates the period of the generator, the number of unique values before the sequence starts to repeat,

```
 1 | def main():
 2 |     m = 10
 3 |     for a in range(1,m):
 4 |         for c in range(m):
 5 |             for s in range(m):
 6 |                 x = s
 7 |                 print "(a,c,s)=(%d,%d,%d): " % (a,c,s),
 8 |                 for i in range(2*m):
 9 |                     x = (a*x+c) % m
10 |                 z = x
11 |                 first = True
12 |                 for i in range(2*m):
13 |                     x = (a*x+c) % m
14 |                     if (x == z) and (first):
15 |                         p = i+1
16 |                         first = False
17 |                     print x,
18 |                 print " (period = %d)" % p
19 |
20 | main()
    | (lcg.py)
```

Lines 3–5 set up a triple loop over a, c, and s. Line 6 copies s to x, it is x that will be updated. Line 7 prints the current set of parameters and starts two loops (lines 8 and 12). We use two loops to ensure that we are in a region of the generator that will be repeating. Note that the loops have $2m$ iterations each. We know the maximum period of this generator is at most 10 ($m = 10$) so we will be sure to be in a region that will repeat. The second loop is used to print out actual values and to track when the repetition starts, this defines the period. This trick works for this simple generator. Lines 9 and 13 are the actual LCG recurrence relation.

If we run this program we will get a long list of output like this,

```
(a,c,s)=(3,9,7): 0 9 6 7 0 9 6 7 0 9 6 7 0 9 6 7 0 9 6 7

                                                    (period=4)
```

telling us that LCG(m, a, c) = LCG(10, 3, 9) with a starting seed value of 7 has a period of 4 out of a possible 10, forever repeating the sequence 0 9 6 7.

Searching the output shows clearly that many combinations of a and c lead to highly ineffective generators with low periods. Figure 2.1 shows this clearly with only 4% of parameters leading to a generator with maximal period.

The only generators with period 10 are those with $a = 1$ and $c \in 1, 3, 7, 9$ regardless of starting seed. Clearly, there is a relationship between m, a, and c which can explain what we are seeing. We should note here that maximal period is desirable but certainly not sufficient. For example, $(a, c) = (1, 1)$ simply delivers the digits 0 through 9 in order starting with the seed plus one.

For an LCG there are certain selection rules which govern the period of the generator. These are given in detail in [2] but we can restate them here as,

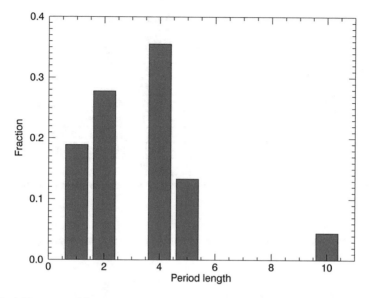

Fig. 2.1 A histogram of the periods for a simple LCG generator with $m = 10$ and all allowed a, c, and seed values. The majority of the generators have a period less than the maximum of 10

1. m and c are relatively prime.
2. $a - 1$ is divisible by the prime factors of m.
3. $a - 1$ is divisible by 4 if m is divisible by 4.

which is satisfied for the four pairs of values (a, c) because $a - 1 = 0$ which is divisible by the prime factors of $m = 10 = 2 \times 5$ and m is not divisible by 4. One case with period 4 is $(a, c) = (8, 9)$. We see that this will not have maximal period because $a - 1 = 7$ which is not divisible by the prime factors of m.

The selection rules are important but for an LCG to be useful we want its period to be as large as possible (ignoring other notions of randomness for the time being). Since we are working with binary computers we would like to fit our generator in a 32 or 64 bit word and we want to use "good" values for the parameters which, beyond maximizing the period, also lead to the passing of as many statistical tests for randomness as possible. To say that a lot of work has been performed in searching for these magic sets of numbers would be an understatement. We will simply list several important sets here, sets that have been used because they are good or, sadly, because they were believed to be good but turned out to be rather poor in the end. Table 2.1 shows these values along with notes about their origin and use.

Three generators in Table 2.1 are of particular importance to us. These are the Apple and C++11 MINSTD generators, which we will discuss below, and the last one in the list, RANDU. Let's look at RANDU first as it is an object lesson in what happens when one has incomplete knowledge.

RANDU is the classic example of a poor LCG that was in wide-spread use starting in the 1960s. Knuth called the generator "truly horrible" [3]. Why is it so

Table 2.1 Parameters for some commonly used and historic linear congruential generators

m	a	c	Notes
2^{32}	1664525	1013904223	Numerical recipes
$2^{31} - 1$	1103515245	12345	glibc (gcc) bits 30..0
2^{32}	214013	2531011	Microsoft Visual C++ bits 30..16
$2^{31} - 1$	16807	0	Apple CarbonLib C++11 mnistd_rand0
$2^{31} - 1$	48271	0	C++11 minstd_rand
$2^{48} - 1$	25214903917	11	Java java.util.Random bits 47..16
2^{32}	69069	1	Vax VMS MTH$RANDOM
2^{31}	65539	0	RANDU

We show the values for RANDU for pedagogical purposes. Do not use this generator for the reasons given in the text

horrible? If we look at it applied two times to a starting value of x_n, noting that $65539 = 2^{16} + 3$, we get,

$$x_{n+2} = (2^{16} + 3)x_{n+1} = (2^{16} + 3)^2 x_n$$

$$= (2^{32} + (6)2^{16} + 9)x_n$$

$$= (6(2^{16} + 3) - 9)x_n$$

$$= 6x_{n+1} - 9x_n$$

where each term is mod 2^{31} and 2^{32} mod $2^{31} = 0$. The last line shows the heart of the problem: each value is trivially related to the previous two. Because of this, if one were to plot triplets from RANDU they would fall neatly into planes in 3D space via the theorem by Marsaglia [4] and as we can see clearly in Fig. 2.2. This is analogous to the two photographs in Fig. 2.3 showing two views of the same stand of trees.

RANDU was widely used in the 1960s into the early 1970s and beyond. Because of this, some have questioned early Monte Carlo results. It should be pointed out, however, that converting RANDU output to 32-bit floating-point by dividing by 2^{31} drops many bits which may help alleviate the problem somewhat. Regardless, RANDU should never be used for actual work.

In 1988, Park and Miller [5] attempted to sort through the confusion of parameters for LCGs and propose a "minimum standard" generator which has been called the MINSTD generator. Their purpose was to find a generator that was good for 32-bit computers and which could be used to bring some consistency to implementations. The generator that they initially recommended was,

$$x_{n+1} = 16807x_n \text{ mod } 2^{31} - 1$$

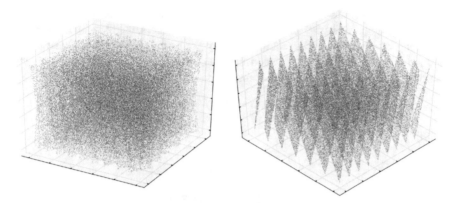

Fig. 2.2 Plotting 100,000 points generated from RANDU as triplets scaled $[0, 1)$. From the viewpoint on the left, the generator looks like it is filling the space, as expected. However, when rotated and viewed as on the right, we see clearly that the points are falling into planes and not filling the 3D space

Fig. 2.3 Two photographs of the same stand of trees illustrating the same effect seen in RANDU in Fig. 2.2. A front view (left) and the side view (right) showing the rows of trees. Photographs by the author

which has a period of approximately $2^{31} \approx 10^9$. This period was likely considered long in 1988 but it is not considered so now. Still, the MINSTD generator is widely used and perfectly adequate for many noncritical uses (e.g. games).

In 1993 the generator was criticized [6] prompting the original authors to respond [6] and simply recommend that $a = 16807$ be replaced with $a = 48271$. In Table 2.1 we see this change between the Apple CarbonLib implementation and the C++11 implementation. The criticism was, in part, related to the theorem of Marsaglia mentioned above and the fact that all LCG generators will ultimately fall into planes when viewed as n-tuples. The multiplier 16807, according to Marsaglia, happens to be a particularly poor one. This criticism was countered by Park and Miller by suggesting $a = 48271$ as a better choice. It should be noted that in their original paper, Park and Miller commented that better constants might be found in the future.

When implementing an LCG generator we need to calculate the product ax_n which, recalling that both a and x_n are likely 32-bit integers, might result in overflow requiring 64-bits to store the result. This may not be feasible or desirable so to get around this problem Schrage introduced a decomposition [7] which calculates the product without overflow. The decomposition is,

$$ax \bmod m = A(x) + B(x)m$$

with,

$$A(x) = a(x \bmod q) - r(x/q)$$

$$B(x) = \begin{cases} 1, & \text{if } A(x) < 0 \\ 0, & \text{otherwise} \end{cases}$$

$$q = m/a$$

$$r = m \bmod a$$

where all division is integer division.

This decomposition is based on the approximate factorization of m via $m = aq + r$ with $q = m/a$ and $r = m \bmod a$. This implies that for $x < m$, which is always the case here, both $a(x \bmod q)$ and $r(x/q)$ will both be less than 32-bits (signed). It is the signed part that adds m back in if the difference of these two quantities is negative.

We can apply Schrage decomposition to define a MINSTD function in C as,

```
1   #define M 2147483647
2   #define A 48271
3   #define Q 44488
4   #define R 3399
5
6   int seed = 1;
7
8   int minstd(void) {
9        seed = A * (seed % Q) - R * (seed / Q);
10       seed += (seed < 0) ? M : 0;
11       return seed;
12  }
13
14  double dminstd(void) {
15       return minstd() / (double)M;
16  }
```
(minstd.c)

where we have used the updated multiplier $a = 48271$ as recommended by Park and Miller [6].

The seed is stored in `seed` and initialized to 1 in line 6. Typical uses initialize this value prior to calling `minstd`. If a fixed value is used, each run of the program will produce the exact same sequence of values. Alternatively, initializing the seed to an external random value, perhaps based on the system time, will produce nondeterministic behavior (only apparently so, of course, depending upon the randomness of the initializer).

The actual calculation of the next value, stored in `seed`, is in line 9. We use precomputed Q and R values for the given M and A. Line 10 checks for the case when we have gone negative and need to add M back in. The function `dminstd` is a convenience function to generate doubles in the range $[0, 1)$ by dividing by M.

If we use this function to generate 20 million doubles (10 million points) and pass these to a slightly modified version of our program above to estimate π we get, for the default seed, $\pi \approx 3.1425808$ which is off by 0.03% to eight places showing that for this particular use case MINSTD is a reasonable option. We will subject MINSTD to additional tests in Chap. 4.

A property of linear congruential generators that can lead to improper use is the periodicity of lower-order bits. When m is a power of two, the modulo operation becomes simple bit masking, no division required. Specifically, for integer values, $x \bmod n = x$ AND n, $n = 2^k$, making using such an m value attractive. However, this leads to periodicity in the lower order bits that is far below that of the generator as a whole. We can see this plainly if we examine a histogram of bit values for specific bits in the output of such a generator.

The Vax generator in Table 2.1 uses $m = 2^{32} = 4294967296$ so it should exhibit this property. We can extract specific bits from the generator with a simple C program,

```
1   #include <stdio.h>
2   #include <stdint.h>
3   #define M 4294967296
4   #define A 69069
5   #define C 1
6   #define B 0
7
8   uint32_t seed = 1;
9   uint32_t randvax(void) {
10      seed = (uint32_t)((A*(uint64_t)seed + C) % M);
11      return seed;
12  }
13
14  void main() {
15      FILE *f;
16      uint8_t b;
17      int i,k;
18      uint32_t s;
19
20      f = fopen("vax_bit.dat", "w");
```

```
21      for (i=0; i < 4000000; i++) {
22          b = 0;
23          for (k=0; k<8; k++) {
24              s = randvax();
25              b <<= 1;
26              b |= (s & (1<<B)) ? 1 : 0;
27          }
28          fwrite((void*)&b, 1, 1, f);
29      }
30      fclose(f);
31 }
```
(lcg_vax.c)

where lines 3 through 5 define the LCG constants and line 6 sets B to the bit number
to store. If we change B we can output the value of any bit from 0 to 31.

Lines 8 through 12 define a function, randvax, that implements the Vax LCG
using a fixed starting seed of 1. For linear congruential generators with m a power
of two, the seed should be an odd number.

The main program in lines 14 through 31 opens an output file (line 20) and loops
4 million times (line 21). The inner loop in lines 23 through 27 makes eight calls to
randvax, each time storing the desired bit of b in b by shifting up one bit position
(lines 24–26). We do this by masking the desired bit of s with 1 shifted up B bits.
The output of this AND operation will be either zero if the bit is not set or 2^B. This
then OR's b with a zero or a one in the lowest bit position. The shift of line 25 will
make room for the next bit. When b contains eight bits it is written to the output file
in line 28.

The output of this program is a file of bytes representing the selected bit values
returned by 32 million calls to randvax. If the output of this LCG is truly random
in all bit positions then regardless of the value of B we should see, for each byte of
the output file, all possible values with roughly equal frequency. However, as might
be expected from doing this exercise in the first place, this will not be the case.
Figure 2.4 shows histograms of the distribution of byte values for bits 0, 1, 2, 4,
6, 8, 10, 12, 14, 16, 24, and 31. Clearly, not all bits of this generator are random.
In fact, the lowest order bits are simply oscillating. For bit 0, the output toggles
between 0 and 1 which means each byte is $55_{16} = 01010101_2$. For bit 1 we get
$CC_{16} = 11001100_2$. In Fig. 2.4 this becomes a single value in the histogram. If we
continue, we see that bit 2 returns $F0_{16} = 11110000_2$ repeating and bit 4 gives
5F 1B A0 E4 repeating. So, as we look at successively higher bits in the output, the
pattern gets longer and longer. Again, we see this in Fig. 2.4 where by bits 24 and
31 we do have all byte values well represented.

We can use the data files with the Monty Hall Dilemma program from Sect. 1.1
as a simple test of their quality. Clearly, very low-order bits will not be useful. What
about bit 10? Doing that gives,

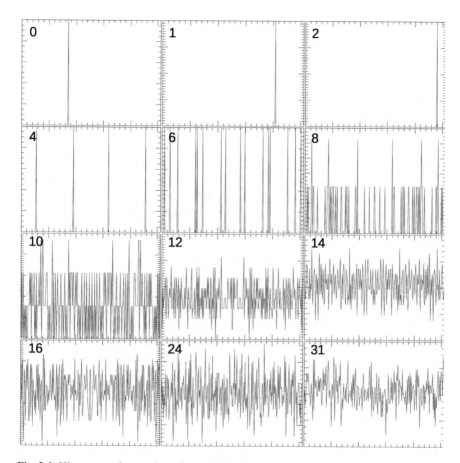

Fig. 2.4 Histograms of output bytes for each bit of the Vax LCG generator

```
Results after 2000000 simulations:

   Wins on initial guess  = 562500 (28.1250 %)
   Wins on changing guess = 1437500 (71.8750 %)
```

which is not acceptable. Bit 16 gives,

```
Results after 2000000 simulations:

   Wins on initial guess  = 671139 (33.5570 %)
   Wins on changing guess = 1328861 (66.4430 %)
```

which is within what we expect to see (recall, changing doors wins 2 out of 3 times).
Finally, moving to bit 24 gives,

```
Results after 2000000 simulations:

  Wins on initial guess  = 666624 (33.3312 %)
  Wins on changing guess = 1333376 (66.6688 %)
```

which is well within our expectations.

Does this mean that this generator is useless? No, but it does mean that care should be taken when using the output of this, or any, linear congruential generator. For example, we would be disappointed if when we wanted random byte values we took the output of the `randvax` function above and used it modulo 256 thereby preserving only the lowest order bits. A good way to use these generators is to divide the output integer value by m using floating-point division to form a floating-point value, $[0, 1)$. Multiplying this value by the maximum we desire, say y, gives an output in the range $[0, y)$. Using floor or ceiling on this value will give us an integer preserving the randomness properties of the generator.

In this section we introduced linear congruential generators, showed how they work, and learned a bit about how to select proper parameter values. We also saw that while good generators exist, we must be careful when using them and aware of the limited period they offer. We implemented the minimum "standard" generator in C and showed the Schrage method for computing using only 32-bit operations. Finally, we used individual bits of a classic LCG to show that we need to use the entire output of the generator in order to get a sequence that will be useful.

2.3   Mersenne Twisters

The Mersenne Twister (MT, or "the twister") is currently the *de facto* standard for pseudorandom number generators, at least for tasks that do not require a cryptographically secure generator. We will take a good look at this generator because it is so widely used, however, the reader is encouraged to take a look at the generators of Sect. 2.4 as these are simpler to implement, provably better, and definitely faster than the Mersenne Twister and its variants.

The mathematics behind the MT are daunting, even for mathematicians. We will not dive so deep but we will at the same time endeavor to impart the framework in which the MT operates. Specifically, we will present C code for the twister and then work through some of the framework behind it. We will test the twister in Chap. 4 along with the other generators we've discussed so far in this book.

The original paper by Matsumoto and Nishimura describing the twister came out in 1997 [10]. The version described in the original paper was improved slightly by the authors a few years later and this is the version we will use here. The improvement was to the initialization routine to correct for initial output that was not very random when the initialization had many zeros in it. The twister was the first fast generator to produce high quality, uniformly distributed, pseudorandom 32-bit numbers. Since its release, the twister has been supported by the authors and has been implemented in numerous programming environments and operating systems. No pseudorandom number generator is perfect, but for general use, the twister is provably quite good. Let's look at two characteristics of the twister which demonstrate its quality: the period and the distribution of output values.

The Mersenne Twister has a period of $2^{19937} - 1$, a Mersenne prime, hence the "Mersenne" part of the name. This is a 6002 digit number that is hard to put into human terms. The MINSTD LCG has a period of just over 2 billion. Two billion seconds is 63.4 years. That is a number we can relate to. Physicists estimate that there are approximately 10^{89} photons forming the Cosmic Microwave Background left over from the Big Bang. This number accounts for the vast majority of photons in the universe. The period of the MT is $(2^{19937} - 1)/10^{89} \approx 4 \times 10^{5912}$ times larger than the number of CMB photons implying that the period of the MT is so large one would need to count all the photons in 4×10^{5912} universes, assuming the theoretical multiverse hypothesis to be true, in order to approximate it. Claude Shannon put a conservative lower bound on the number of possible chess games at 10^{120} [8]. While this number is 31 orders of magnitude larger than the estimated number of CMB photons it still leaves us with a 5882 digit number when dividing the twister's period.

Assuming a fair coin, the probability of flipping n heads in a row is $1/2^n$. This means that the reciprocal of the period of the MT is like flipping 19,937 heads in a row. Intuitively, we know that this is very, very, very unlikely to ever happen. Perhaps this is the best we can do to put the period of the Mersenne Twister into human terms. We will see generators below with periods far longer than this.

As we stated above, the period of a pseudorandom generator is very important, but it isn't the only consideration. We saw linear congruential generators with maximal period that were perfectly predictable. So, not only do we need a large period, we need to fulfill our definition of a random sequence from Chap. 1. One way to characterize this unpredictability is to consider how well the generator fills space. We want the generator to fill space equally for some dimensionality, k. This is known as *equidistribution* to k dimensions.

Specifically, we want to find the k for which this equal filling is true for the leading v bits of the output of a generator. This is a little confusing because in reality k is a function of v so we really want to find a set of k values, for each v less than or equal to the bit width of the generator, $1 \leq v \leq 32$. For example, it is known that the twister is 623-dimensional equidistributed for $v = 32$ bits.

How do we best interpret this statement? Consider taking 623 successive outputs of the generator. Each one of these is a 32-bit integer which we express in binary. We then form a $623 \times 32 = 19936$ bit vector by concatenating each 32-bit pattern together. If we do this for each successive set of 623 outputs from the generator over a full period, $2^{19937} - 1$, we will find that each of the 2^{19936} 19936-bit patterns appears the same number of times. The exception is the all zero pattern which will appear one fewer time.

Another way to understand k-distribution is geometrically. To do this, form the vectors as before, of k dimensions (623 for the twister), and divide each component by the maximum output (2^{32}) to form values in the range $[0, 1)$. These are then the coordinates of a point in a k dimensional unit hypercube. If we divide each axis into 2^v bins we will partition the hypercube into subcubes. Then, if we count the number of points in each subcube, over the entire period of the generator, and each count is the same in each cube, then we can say that the generator is equidistributed in k dimensions.

Naturally, we cannot do this in a visual way for the twister but we can quite easily get a feel for how well it is filling space by fixing $k = 3$ and looking at various v values for a large number of samples from the generator. Fortunately for us, the NumPy library for Python implements the twister so we can easily setup a program which draws triplets from the generator and fills in a 3D histogram (because $k = 3$) for various v values. We can use this to gain some insight to the claim that the twister is equidistributed for all k up to 623 at $v = 32$ bits.

The Python program is straightforward,

```
 1   import numpy as np
 2   import cPickle
 3
 4   def main():
 5       results = {}
 6       N = 100000000
 7
 8       for v in range(1,6):
 9           h = np.zeros((2**v,2**v,2**v), dtype="uint32")
10           for i in xrange(N):
11               n = np.random.random(3)
12               x0 = np.floor(n[0] / (2.0**(-v)))
13               x1 = np.floor(n[1] / (2.0**(-v)))
14               x2 = np.floor(n[2] / (2.0**(-v)))
15               h[x0,x1,x2] += 1
16           results[v] = h.reshape(2**v*2**v*2**v)
17
18       cPickle.dump(results, open("kdist.pkl","w"))
     (kdist.py)
```

where we loop over v (line 8) and for each v create a histogram, h, in line 9. We then draw $3N$ samples from the twister in groups of three (lines 10 and 11). Lines 12–14 determine which subcube of our unit cube we have fallen into. Since $k = 3$ the histogram has three dimensions. There are 2^v bins along each axis. Line 15 increments the counter for this subcube. When the histogram is complete we store it (line 16) and when all v values are complete we store the entire set of histograms (line 18).

Running this program produces an output file in Python pickle format. This is simply for convenience. To analyze the results we examine each histogram for $v = 1\ldots5$. The histograms are of different sizes, each dimension is 2^v bins, but we can summarize them simply by looking at some basic statistics like the mean,

Table 2.2 Mean, standard
deviation, and coefficient of
variation for the number of
points ($k = 3$) in each
subcube of the unit cube for
different v values

v	Mean (μ)	Std dev (σ)	CV (σ/μ)
1	12500000	2171	0.000174
2	1562500	1162	0.000744
3	195313	460	0.002358
4	24414	156	0.006380
5	3052	56	0.018210

If we ran the program over the entire period
the number in each bin would approach a
constant value

standard deviation and the coefficient of variation (the standard deviation divided by the mean). These values are in Table 2.2. We see that for 300000000 draws from the twister we have for small v values some intuition that the number of values in each subcube is approaching a constant ($\sigma = 0$).

The canonical C code for the twister is given on the MT website [9]. There you will find the most recent version. Thanks to a liberal license we can reproduce the core code here as shown in Fig. 2.5.

In Fig. 2.5 two functions, `init_genrand` and `genrand_int32`, are all that is needed to generate unsigned integers in the range [0, 0xFFFFFFFF]. First, one calls `init_genrand` with an unsigned 32-bit seed value, then subsequent calls to `genrand_int32` will return the integers. Line 25 checks to see if `init_genrand` has been called and if not, supplies a seed value of 5489. One thing to note about the MT code is that except for the initialization step (line 14) there are no multiplications when generating a pseudorandom integer. Compare this to the multiplication necessary for an LCG. While this was an advantage in the late 1990s when the twister was introduced, it is less so now.

As before, if we want floating-point output, we can define a simple function to calculate a double in the range [0, 1),

```
1  double genrand_real2(void) {
2      return genrand_int32()*(1.0/4294967296.0);
3  }
```
(mt19937ar.c)

where we simply divide the 32-bit integer by 2^{32}.

This division implies that the resolution of the floating-point number is about 2^{-32}. Or, to put it another way, that we are partitioning the interval [0, 1) into $2^{32} - 1$ steps and selecting one of them. While this is likely adequate for the vast majority of cases we can do better. Why? Because the IEEE 754 standard dictates that a double precision floating-point number have 53 bits in the significand. Since the simple function only gives $2^{32} - 1$ possible output values we are not making maximum use of the bits available to us. We can fix this by defining a second function to return a floating-point number but one that uses all the bits of the significand. Fortunately, the current canonical version of the twister includes just such a function,

```
 1   #define N 624
 2   #define M 397
 3   #define MATRIX_A 0x9908b0dfUL
 4   #define UPPER_MASK 0x80000000UL
 5   #define LOWER_MASK 0x7fffffffUL
 6
 7   static unsigned long mt[N];
 8   static int mti=N+1;
 9
10   void init_genrand(unsigned long s) {
11       mt[0]= s & 0xffffffffUL;
12       for (mti=1; mti<N; mti++) {
13           mt[mti] =
14           (1812433253UL * (mt[mti-1] ^ (mt[mti-1] >> 30)) + mti);
15           mt[mti] &= 0xffffffffUL;
16       }
17   }
18
19   unsigned long genrand_int32(void) {
20       unsigned long y;
21       static unsigned long mag01[2]={0x0UL, MATRIX_A};
22
23       if (mti >= N) {
24           int kk;
25           if (mti == N+1)
26               init_genrand(5489UL);
27           for (kk=0;kk<N-M;kk++) {
28               y = (mt[kk]&UPPER_MASK)|(mt[kk+1]&LOWER_MASK);
29               mt[kk] = mt[kk+M] ^ (y >> 1) ^ mag01[y & 0x1UL];
30           }
31           for (;kk<N-1;kk++) {
32               y = (mt[kk]&UPPER_MASK)|(mt[kk+1]&LOWER_MASK);
33               mt[kk] = mt[kk+(M-N)] ^ (y >> 1) ^ mag01[y & 0x1UL];
34           }
35           y = (mt[N-1]&UPPER_MASK)|(mt[0]&LOWER_MASK);
36           mt[N-1] = mt[M-1] ^ (y >> 1) ^ mag01[y & 0x1UL];
37           mti = 0;
38       }
39
40       y = mt[mti++];
41       y ^= (y >> 11);
42       y ^= (y << 7) & 0x9d2c5680UL;
43       y ^= (y << 15) & 0xefc60000UL;
44       y ^= (y >> 18);
45       return y;
46   }
```
(mt19937ar.c)

Fig. 2.5 The Mersenne Twister in C

```
1  double genrand_res53(void) {
2      unsigned long a=genrand_int32()>>5,
            b=genrand_int32()>>6;
3      return(a*67108864.0+b)*(1.0/9007199254740992.0);
4  }
```
(mt19937ar.c)

We make two calls to the twister (line 2) to get two 32-bit integers or 64 random bits. For the first we shift down by 5 bits and for the second we shift down by 6 bits. This drops 11 of the 64 bits leaving us with 53 bits. Line 3 calculates the double precision floating-point number based on these 53 bits. This gives us the full range of values expressible by a double precision float in the range [0, 1). Again, while seldom necessary, it is there and useful when extreme resolution is desirable (but at the expense of two calls to the core code). It is important to recall that the two floating-point functions above are also part of the core twister code.

We can easily use the MT code to generate a file of random bytes. We can then pass these bytes to our estimate of π program above and to our Monty Hall Dilemma program from Sect. 1.1 to see what sort of output we get.

The program to generate the byte file is straightforward,

```
1   void main() {
2       FILE *f;
3       unsigned long b;
4       int i;
5
6       f = fopen("mttest_bytes.dat","w");
7       for(i=0; i<10000000; i++) {
8           b = genrand_int32();
9           fwrite((void *) &b, sizeof(unsigned long), 1, f);
10      }
11      fclose(f);
12  }
```
(mttest.c)

where we create an output file (line 6) and make 10,000,000 calls to the twister (line 8) writing the 32-bit number to disk each time (line 9). This creates a file with 40,000,000 random bytes. Estimating π with this output file, using 2 bytes at a time, gives $\pi \approx 3.1416920$, which is quite reasonable.

Using this file with the Monty Hall program gives,

```
Results after 20000000 simulations:

Wins on initial guess  = 6665058 (33.3253 %)
Wins on changing guess = 13334942 (66.6747 %)
```

which is exactly as expected.

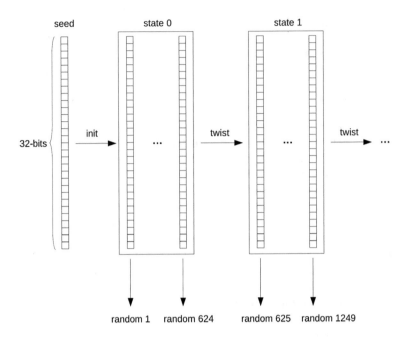

Fig. 2.6 The essential operation of the Mersenne Twister. An initial seed value is used to generate the initial state 0. The state consists of 624 32-bit values which supply the first 624 pseudorandom numbers. When a state is exhausted the twist happens to move to the next state which supplies another 624 32-bit values. This continues until no more values are needed. The output repeats after $2^{19937} - 1$ values. Graphic modeled after the drawing by David Wong found here: https://www.cryptologie.net/article/331/how-does-the-mersennes-twister-work/

By now most readers are starting to wonder exactly what the twister is doing. We typically describe code in this book in some detail but the cryptic nature of the C code above requires a different approach in this case.

Most pseudorandom generators have a concept of a state. For the LCG the state is a single 32-bit number. For the twister the state is 624 words where each word is a 32-bit value (`unsigned long`). The algorithm pulls unsigned 32-bit values from this state until they are exhausted. When exhausted the "twist" happens which generates a new state of 624 values. This continues as long as values are needed. The essential operation of the twister is shown graphically in Fig. 2.6.

Figure 2.6 shows an initial 32-bit seed used to set up the initial state of the twister. This corresponds to the function `init_genrand` (lines 10–17). The second function, `genrand_int32`, returns one of the 32-bit values in the current state (line 40) first multiplying it, thinking of the 32-bit value as a vector with elements in {0, 1}, by a tempering matrix (lines 41–45).

Wait you may say, what tempering matrix? The matrix is implicit in the operations of lines 41 through 44 in Fig. 2.5. By performing the given bitwise shifts and other operations the net effect is to multiply y by such a matrix which in [10] is called T, a 32×32 matrix. The purpose of these lines is to improve the k-distribution to v-bit accuracy.

Lines 23 through 38 of genrand_int32 are only run when the current state is exhausted meaning 624 values have been requested (the counter is in mti). This code implements the twist to move to the next state. Lines 25–26 check to see if genrand_int32 has been called before init_genrand and if so, initializes the generator with a fixed seed of 5489. The actual twist is in lines 27–37.

The Mersenne twister depends upon a recurrence relation,

$$\mathbf{x}_{k+n} \leftarrow \mathbf{x}_{k+m} \oplus (\mathbf{x}_k^u | \mathbf{x}_{k+1}^l)A, \; k = 0, 1, 2, \ldots$$

where $\mathbf{x}$ is a 32-bit binary number. From a theoretical point of view $\mathbf{x}$ is really a $w = 32$ dimensional row vector in $GF(2)$, the Galois field with two elements, $\{0, 1\}$. Thinking of it as a binary number works, too. It is not necessary to sort through the recurrence directly, we can examine the code in Fig. 2.5 to see in practice the relation becomes,

```
y   =  (mt[kk]&UPPER_MASK) | (mt[kk+1]&LOWER_MASK);
mt[kk]  =  mt[kk+M] ^ (y >> 1) ^ mag01[y & 0x1UL];
```

where the next value, k (kk), is stored in the array mt representing the state. The first line sets up y as the UPPER_MASK bits of the current value OR-ed (concatenated) with the lower LOWER_MASK of the next value in the old state. Note, the code updates the state in place hence referencing mt[kk] on the RHS of the top line and setting it on LHS of the next line. The mask values are in lines 4 and 5 of Fig. 2.5 and keep the top-most bit of mt[kk] and OR in the lower 31 bits of mt[kk+1].

The second line above is even more cryptic than the first. Looking at the relation we see that the expression,

```
(y >> 1) ^ mag01[y & 0x1UL]
```

corresponds to

$$(\mathbf{x}_k^u | \mathbf{x}_{k+1}^l)A$$

where the matrix A is a 32×32 matrix of a form that makes multiplication by the 32-bit row vector $\mathbf{x}$ very simple. The A matrix is mostly diagonal so that in the end the multiplication is simply a multiplication of two vectors. Hence the definition of MATRIX_A in line 3 of Fig. 2.5. What multiplication? The operations here are modulo 2 because they really take place in GF(2) so bitwise AND is multiplication and bitwise XOR is addition. This explains the phrase mt[kk+M] ^ ... which corresponds to $\mathbf{x}_{k+m} \oplus \ldots$.

The loops in lines 27 through 36 of Fig. 2.5 shift the entire 624 element array of 32-bit values according to the constants N=624 and M=397. The N value comes from $\lfloor 19937/32 \rfloor + 1 = 624$ and M is chosen, along with MATRIX_A, to give the twister a maximum period. The values for the tempering matrix (lines 41–44) are similarly chosen to maximize the equidistribution properties of the twister. See Table II in [10] for 19937.

The only part of the code we have not discussed is the initialization routine,

```
10  void init_genrand(unsigned long s) {
11      mt[0]= s & 0xffffffffUL;
12      for (mti=1; mti<N; mti++) {
13          mt[mti] =
14          (1812433253UL * (mt[mti-1] ^ (mt[mti-1] >> 30))
                + mti);
15          mt[mti] &= 0xffffffffUL;
16      }
```

which takes the given 32-bit seed value (s) and puts it into the state table as
the first entry (line 11). The mask simply ensures that only the first 32 bits of
s are used if the type unsigned long happens to be larger than 32-bits. The
remaining 623 elements of the initial state are calculated from previous element
via the mystical line,

```
mt[mti] = (1812433253UL*(mt[mti-1] ^ (mt[mti-1] >> 30)) + mti)
```

where the constant 1812433253 is the Borosh-Neiderreiter multiplier for a multiplicative LCG with 2^{32} as the modulus. In effect, the initialization is similar to an
LCG and ensures that poor initialization conditions are not encountered. Line 15
simply masks the table values to ensure only 32-bits are used.

The full Mersenne Twister package available from [9] includes a function
excluded above to initialize the generator from an array of 32-bit values instead
of from a single 32-bit seed value. This enables very fine-grained control of the
generator's initial state beyond the 2^{32} possibilities of the simple seed approach.

The C code given above is the original code, corrected for the initialization issue
discovered later, but it only produces 32-bit integers. There is a faster variant of the
twister available in the SFMT package, also found on the MT website [9]. SFMT
stands for "SIMD-oriented Fast Mersenne Twister" and this version will work best
on modern pipelining processors. It was developed by Mutsuo Saito and Makoto
Matsumoto in 2006 [11].

It is claimed that the SFMT code is about twice as fast as the original twister.
We will test this claim below. The SFMT code itself provides a simple API
for generating 32-bit and 64-bit pseudorandom numbers along with floating-point
versions. The API also allows for other exponents to implement twisters with
periods other than the original $2^{19937} - 1$. In each case the period is a Mersenne
prime from a low of $2^{607} - 1$ to an incredible $2^{216091} - 1$ which is a number of
65,050 digits in base 10.

Let's compare the runtime of the original twister with SFMT. We will also
test against the MINSTD code given above. In all cases we are timing, via
the clock_gettime system routine, with CLOCK_MONOTONIC, so that we have
precise timing without gaps. The task is simple, generate 1 billion (10^9) 32-bit
pseudorandom numbers. The compiler is gcc.

If we simply compile at different optimization levels (-O1 and -O3) we get as the mean $\pm$ SE over ten runs,

Generator	No opt (s)	$-$O1 (s)	$-$O3 (s)
SFMT	14.5398 $\pm$ 0.0076	5.77167 $\pm$ 0.0050	4.67965 $\pm$ 0.0042
MT	10.6487 $\pm$ 0.0156	5.05582 $\pm$ 0.0052	5.90670 $\pm$ 0.0043
MINSTD	10.5273 $\pm$ 0.0065	8.19713 $\pm$ 0.0075	6.80746 $\pm$ 0.0045

where it is clear that compiler optimization makes a rather large difference in this case. Without it, SFMT is by far the slowest of the generators, not the fastest. However, with sufficient optimization, the -O3 case, the SFMT generator becomes the fastest. This is promising. While it is not clear which of the many optimizations of the gcc compiler cause the code to perform so well, it tracks with the claims in [11] that the generator works best with modern processors. Surprisingly, and it is not clear why since the optimizations are cumulative, going from -O1 to -O3, the original Mersenne twister *increases* its run time while the other two generators continue to run faster and faster.

For the test above we used 32-bit Linux running on a quad core Intel i5-2310 processor. If we repeat the test using a single core Intel Atom N270 processor we get,

Generator	No opt (s)	$-$O1 (s)	$-$O3 (s)
SFMT	78.6056 $\pm$ 0.3004	20.6349 $\pm$ 0.4054	18.3866 $\pm$ 0.4101
MT	44.0061 $\pm$ 0.5128	29.7715 $\pm$ 0.2849	23.9326 $\pm$ 0.0017
MINSTD	38.3029 $\pm$ 0.1509	21.7989 $\pm$ 0.1389	17.9966 $\pm$ 0.1526

where the run times are naturally longer but we see something else of interest. The MINSTD generator is always the fastest regardless of optimization level, however, the difference between SFMT and MINSTD for -O3 is not statistically significant ($p = 0.39$).

Finally a Raspberry Pi 3 Model B gives,

Generator	No opt (s)	$-$O1 (s)	$-$O3 (s)
SFMT	126.9645 $\pm$ 0.1230	30.2647 $\pm$ 0.0261	21.6074 $\pm$ 0.0216
MT	97.9179 $\pm$ 0.0866	39.6251 $\pm$ 0.0333	34.3268 $\pm$ 0.0327
MINSTD	62.3213 $\pm$ 0.0515	25.0572 $\pm$ 0.0205	11.7068 $\pm$ 0.0119

where for -O3 the nearly factor of two difference between MINSTD and SFMT is highly statistically significant ($p < 0.00001$). Undoubtedly, this is due to low-level architectural differences between the Intel processors and the quad core ARM used by the Raspberry Pi. It may also be due to the size of the table used by the twister which may not fit well in the cache of the ARM processor.

The three tests above indicate that one should never simply accept statements about a particular piece of code or algorithm since, as in speech, context matters. For small systems it seems that simpler generators might be appropriate while for more powerful processors SFMT is a viable alternative to the original twister.

2.4   Xorshift and Variants

The xorshift generator of Marsaglia [12] is a fast, simple to implement, generator with high period. In this section we describe the generator and several of its newer variants that are capable of passing the most stringent of statistical tests (see Chap. 4). In reality, xorshift is not a single generator but an entire family of generators based on triplets of parameters.

The C code for an xorshift generator with period $2^{32} - 1$ is,

```
 1  #define A 13
 2  #define B 17
 3  #define C  5
 4  unsigned long s32 = 2463534242UL;
 5
 6  unsigned long xorshift32(void) {
 7       s32 ^= (s32<<A);
 8       s32 ^= (s32>>B);
 9       s32 ^= (s32<<C);
10       return s32;
11  }
```
(xorshift32.c)

where A, B, and C are three parameters that select which of the family of generators we will be using. See [12] for tables of parameters. Like MINSTD, this generator uses a single 32-bit unsigned integer as the seed. Any value that is not zero is acceptable.

The code could hardly be simpler, just three XOR statements each XOR-ing the seed with a shifted version of itself. We will see below why this could be expected to produce a pseudorandom sequence. Once the seed has been moved to the next state, simply return it (line 10). Of course, as before, we can turn the unsigned integer into a floating-point number by dividing by 2^{32} or by using the function genrand_res53 above in Sect. 2.3 to use all 53-bits of precision available in an IEEE 754 double.

Increasing the state space to four 32-bit values enables a straightforward extension to a generator with period $2^{128} - 1$,

```
 1  unsigned long s128[] = {123456789,362436069,
    521288629,88675123};
 2  #define A128 11
 3  #define B128  8
 4  #define C128 19
 5
 6  unsigned long xorshift128(void) {
 7       unsigned long t;
 8       t = (s128[0]^(s128[0]<<A128));
 9       s128[0]=s128[1]; s128[1]=s128[2]; s128[2]=s128[3];
```

```
10|     s128[3] = (s128[3] ^ (s128[3]>>C128)) ^ (t^(t>>B128));
11|     return s128[3];
12| }
```
   (xorshift128.c)

where the seed is in `s128`. Note, we only return the last 32-bit value, we could return any of the four values. For that matter, one could look at all four together to treat the seed as a $32 \times 4 = 128$ bit number. We use `A128`, `B128`, and `C128` as the parameters of this particular generator. The seed values cannot all be zero. The values given for the parameters and seed are taken directly from [12]. Notice that `xorshift128` has the three steps of `xorshift32` in lines 8 and 10. Line 8 XOR's the first seed value with a shifted version of itself and then keeps it for later use. Line 9 simply moves the other three 32-bit see values into the next position. Finally, line 10 combines lines 8 and 9 of `xorshift32` into one step and updates the last seed value.

If we run these generators on our well-worn Monty Hall Dilemma program we get $(33.3279\%, 66.6721\%)$ for xorshift32 and $(33.3312\%, 66.6688\%)$ for xorshift128 simulating 20,000,000 games. These are the expected percentage wins for the initial guess and changing doors so we have some confidence that the generators are working properly.

How can something so simple work so well? In order to understand what the generator is doing we need to learn a little bit of abstract algebra. The generator works in a field, specifically, in $\mathscr{F}_2$. A field is a group that supports two operations. A group is a set of symbols and an operation on them that meets a certain definition. For our purposes, the two operations are addition and multiplication on the set of binary digits, $\{0, 1\}$. In this set, the times table consists of four entries,

$$0 \times 0 = 0$$
$$0 \times 1 = 0$$
$$1 \times 0 = 0$$
$$1 \times 1 = 1$$

which is a logical AND. So, in $\mathscr{F}_2$, multiplication is AND. What about addition? The addition table is,

$$0 + 0 = 0$$
$$0 + 1 = 1$$
$$1 + 0 = 1$$
$$1 + 1 = 0$$

which we recognize as the XOR operation. So, if we are working in $\mathscr{F}_2$, we only need the bit operations of AND for multiplication and XOR for addition. This is a good thing.

In [12] we see that a pseudorandom number generator is some operation, $f()$, on some state, z. Repeated applications of this operation generate the sequence of pseudorandom values. If z_0 is the starting state, in this case a vector in $\mathscr{F}_2$ which, conveniently, is really just a binary number, each bit an element of the vector, then $z_1 = f(z_0)$ and $z_2 = f(z_1) = f(f(z_0))$ and so on.

We see that things are falling into place. We are operating on bits ($\mathscr{F}_2$) and can use AND and XOR to implement our operations. The only thing missing is a good choice of $f()$. Marsaglia presents a theorem in [12] stating that, for an n-bit state vector, z, there exist binary matrices, T, $n \times n$, which when applied repeatedly to z generate all possible n-bit vectors. Therefore, if we find a good T matrix, we know that for any z_0 (an n-bit binary vector), the sequence $zT, zT^2, zT^3, \ldots$ will ultimately generate all possible n-bit z values. This is exactly what we want. However, what is a good T? There are two criteria for a good T matrix. First, it must be efficient to generate zT otherwise the generator will be too slow to be useful and, second, it must be of the proper sort to cause all n-bit values to be generated.

Marsaglia shows that the second criteria is satisfied when the order of T is $2^n - 1$. To satisfy the first criteria we want a form that makes the potentially very expensive matrix multiplication almost trivial. It turns out that there is such a form. Let L be an $n \times n$ matrix of all zeros except for the subdiagonal, which is all ones. Then, it can be shown that zL is a left shift by one bit position. For example, let z be a binary row vector, $z = \{1, 0, 1, 1, 0, 1\}$. Then, since we have six dimensions, L is,

$$
L = \begin{bmatrix}
0 & 0 & 0 & 0 & 0 & 0 \\
1 & 0 & 0 & 0 & 0 & 0 \\
0 & 1 & 0 & 0 & 0 & 0 \\
0 & 0 & 1 & 0 & 0 & 0 \\
0 & 0 & 0 & 1 & 0 & 0 \\
0 & 0 & 0 & 0 & 1 & 0
\end{bmatrix}
$$

which is ones along the subdiagonal. If we then form $z' = zL$ we get the row vector $z' = \{0, 1, 1, 0, 1, 0\}$ which we can easily see is z shifted left one position. If we form $z' = zL^2$ we get two shifts, $z' = \{1, 1, 0, 1, 0, 0\}$. Therefore, we can represent zL^a for some integer, a, as an a-bit left shift or, in C, z << a. If we define $R = L^T$, i.e., the transpose of L, we will get a right shift by one bit instead of a left shift.

With these we can finally form the T matrix we need. It is,

$$
T = (I + L^a)(I + R^b)(I + L^c)
$$

for suitable choices of integers a, b, and c and $n \times n$ identity matrix, I. In [12] many choices are given for $n = 32$ and $n = 64$. Any of these lead to a good generator. In the code above for xorshift32 we used the set Marsaglia mentions in the paper, $(a, b, c) = (13, 17, 5)$, which he called, for unspecified reasons, a personal favorite.

Let's look at forming zT which will lead us directly to the code used in xorshift32. The first part is,

$$
z(I + L^a) = zI + zL^a
$$

where we know that zL^a is z << a. We also know that $z = zI$. The addition is XOR, so the entire expression becomes z ^ (z << a) in code. This is line 7 of xorshift32 above where $z \leftarrow zI + zL^a$. Then lines 8 and 9 continue for multiplication by $I + R^b$ and $I + L^c$. We see now why the algorithm is called "xor shift" as that is exactly the set of operations necessary to form $z \leftarrow zT$ to move from one state to the next. Conveniently, the state z is an unsigned integer of $n = 32$ bits so we need do nothing more to get the output value.

We can easily compare the runtime of xorshift32, the Mersenne twister, and MINSTD with code like this,

```
1   clock_gettime(CLOCK_MONOTONIC, &start);
2   for(i=0; i < N; i++)
3       b = xorshift32();
4   clock_gettime(CLOCK_MONOTONIC, &end);
5   d = (end.tv_sec + end.tv_nsec*1e-9) -
6       (start.tv_sec + start.tv_nsec*1e-9);
```

If we run this code ten times for these generators ($N = 100,000,000$) we get,

	mean$\pm$SE
MINSTD	0.680206 ± 0.000511
Mersenne Twister	0.587295 ± 0.000430
xorshift32	0.295628 ± 0.001346

where we are showing the mean$\pm$SE for times in seconds. It is clear that xorshift32 is on average more than twice as fast as the Mersenne Twister and MINSTD. Note that we used gcc and compiled with -O3.

Several variants of xorshift have appeared since the original paper was written in 2003. The very latest versions, based on the most widely used variants, are found on Vigna's website [13]. These are xorshift128+ and the latest recommended replacement for xorshift128+, xoroshiro128+. Both have a period of $2^{128} - 1$ so they are similar to the 128-bit version of xorshift above. We will also look at xorshift1024* which has a much larger period of $2^{1024} - 1$. The Vigna generators include jump functions which we will use in Chap. 5 on parallel generators. The code for these generators is in the public domain and the listings below are exactly as supplied on Vigna's website. The generators themselves are discussed in [14].

The xorshift128+ generator is currently used in the JavaScript engine for several web browsers including Chrome, Firefox, Safari and Edge. It is also the library generator used by the Erlang programming language [13]. This is high praise. Let's look at the code,

```
 1   uint64_t s[2];
 2   uint64_t next(void) {
 3       uint64_t s1 = s[0];
 4       const uint64_t s0 = s[1];
 5       const uint64_t result = s0 + s1;
 6       s[0] = s0;
 7       s1 ^= s1 << 23; // a
 8       s[1] = s1 ^ s0 ^ (s1 >> 18) ^ (s0 >> 5); // b, c
 9       return result;
10   }
```
(xorshift128plus.c)

where there are two seed values (s) which must not be all bits zero. The function next returns the next 64-bit unsigned integer. Note that we are now using 64-bit values.

This generator, like the original xorshift, relies on three parameters that are marked in the code with brief comments. This generator adds the seed values first to get result (line 5) and then updates them before returning.

As of this writing, however, Vigna recommends replacing xorshift128+ with xoroshiro128+ which also has a period of $2^{128} - 1$,

```
 1   uint64_t s[2];
 2
 3   static inline uint64_t rotl(const uint64_t x, int k) {
 4       return (x << k) | (x >> (64 - k));
 5   }
 6
 7   uint64_t next(void) {
 8       const uint64_t s0 = s[0];
 9       uint64_t s1 = s[1];
10       const uint64_t result = s0 + s1;
11
12       s1 ^= s0;
13       s[0] = rotl(s0, 55) ^ s1 ^ (s1 << 14); // a, b
14       s[1] = rotl(s1, 36); // c
15
16       return result;
17   }
```
(xoroshiro128plus.c)

These three functions, xorshift128, xorshift128+, and xoroshiro128+, are only subtly different from each other. If we consider runtime performance they are also very similar (gcc with -O3 optimization, 64-bit version of xorshift128, 100 million samples),

	mean±SE
xorshift128	0.441591± 0.001616
xorshift128+	0.400676± 0.000278
xoroshiro128+	0.419130± 0.000197

where we see that the original xorshift128 is the slowest. The difference between xorshift128+ and xoroshiro128+ is small but statistically significant. This is of interest as the claim about xoroshiro128+ is that it is faster.

The xorshift128+ and xoroshiro128+ generators introduce a new twist. All the generators we have looked at up to this point return 32-bit values. Therefore, conversion to double precision floats was straightforward as double precision floats have 52 bits in their significant (plus one implied). However, these generators return a random *64-bit* integer. How do we get this to be well represented as a double precision float?

The answer is to explicitly build the double using the 52 most significant bits of the 64-bit pseudorandom integer. We use the 52 most significant bits to minimize the influence of any poor randomization of the lower order bits. If we look at the format of an IEEE 754 double (binary64) number, we see the bits are stored as,

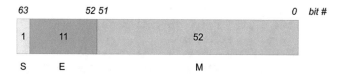

to represent the value as,

$$\pm 1.b_{51}b_{50}b_{49}\ldots b_0 \times 2^e$$

with $b_{51}\ldots b_0$ binary digits and the exponent, e, stored as $e = n - 1023$ with n representing the bits as an unsigned 11-bit integer. The IEEE floating-point format, along with many other number formats, are described in detail in [15].

Therefore, if we wish to construct a bit pattern to represent a double we need to supply the 52 bits of the significand as the lowest 52 bitsi, set the sign bit to zero, and set the exponent to 1023. Why 1023? Because IEEE 754 stores exponents using an offset of 1023. We want the actual exponent to be zero because $2^0 = 1$. If we do all of this, we will have a double precision floating-point number in the range [1.0, 2.0). This is because of the implied leading one bit of the significand. Therefore, we need to subtract 1.0 to get a number in the range [0, 1).

We can define a function to do this for each of the xor generators returning a 64-bit value. For example, for xorshift128+ we can define,

```
1  double dxorshift128plus(void) {
2      uint64_t n = (xorshift128plus()>>12)|0x3FF0000000000000;
3      return *(double *)&n - 1.0;
4  }
```

where line 2 calls the base generator to get an unsigned 64-bit pseudorandom value which is then shifted down 12 bits to preserve the top-most 52 bits. This sets the highest bits of the result to zero which is what we want as these are the sign and exponent bits. If we OR with the given constant we set the most significant bit to zero (positive) and the exponent to 1023 so that the exponent of the resulting double is $2^0 = 1$. We store this carefully crafted bit pattern in n. Line 3 reinterprets the bits of n as a double precision floating-point value which will be in the range [1.0, 2.0). Finally, we subtract 1.0 from this value to shift the range to [0, 1). Note, this approach uses the maximum number of bits in the double precision number which is something we could not do with a single 32-bit integer. The 32-bit integer case required us to make two calls to the underlying generator in order to fill all the bits of the double precision float.

Our final xorshift variant is Vigna's xorshift1024* which has a period of $2^{1024} - 1$. The code is,

```
 1   uint64_t s[16];
 2   int p;
 3
 4   uint64_t next(void) {
 5       const uint64_t s0 = s[p];
 6       uint64_t s1 = s[p = (p + 1) & 15];
 7       s1 ^= s1 << 31; // a
 8       s[p] = s1 ^ s0 ^ (s1 >> 11) ^ (s0 >> 30); // b,c
 9       return s[p] * UINT64_C(1181783497276652981);
10   }
```
(xorshift1024star.c)

where we need to include stdint.h or inttypes.h.

This generator has a long period and passes difficult statistical tests. It is, as expected with the larger state, a bit slower than the generators above. The runtime on the same machine as used above is 0.474718±0.001645 s averaged over ten runs. The constants referred to in the code comments above are specifically searched for as recounted in [14]. The multiplication in line 9 is used to scramble the bits of the value returned without altering the period of the generator since the state (s) is unaffected. This generator should be considered a viable alternative to the Mersenne Twister. In decimal, the period is a 309 digit number that is considerably larger than the 10^{89} value we used above for the estimated number of photons in the universe. And, as with generators of this type regardless of the period, it is equidistributed in all bits, like the Mersenne Twister.

Before moving on we will take a small, fun break. The xorshift family of generators are so simple and have so little overhead that they lend themselves to use on devices with very limited memory. To that end, we will implement a 32-bit xorshift subroutine for a classic 8-bit microprocessor, the 6502. Our demonstration will be tailored for the Apple II line of computers making execution straightforward using one of the many Apple II emulators in the wild. For example, we used AppleWin which runs on Linux via wine. The emulator is available on GitHub (https://github.com/AppleWin/AppleWin). We also tested the code on an actual Apple II computer.

The 6502 code is a direct port of the basic xorshift generator using unsigned 32-bit values. Using the as6502 assembler available on the website associated with this book, we can write the generator as,

```
 1          org $300            42          jsr merge
 2  seed                        43          rts
 3          .byte 01            44  copy
 4          .byte 00            45          lda seed+0
 5          .byte 00            46          sta tmp+0
 6          .byte 00            47          lda seed+1
 7  tmp                         48          sta tmp+1
 8          .byte 00            49          lda seed+2
 9          .byte 00            50          sta tmp+2
10          .byte 00            51          lda seed+3
11          .byte 00            52          sta tmp+3
12  x32                         53          rts
13          jsr step1           54  merge
14          jsr step2           55          lda tmp+0
15          jsr step3           56          eor seed+0
16          rts                 57          sta seed+0
17  step1                       58          lda tmp+1
18          jsr copy            59          eor seed+1
19          ldx #13             60          sta seed+1
20  loop1                       61          lda tmp+2
21          jsr up              62          eor seed+2
22          dex                 63          sta seed+2
23          bne loop1           64          lda tmp+3
24          jsr merge           65          eor seed+3
25          rts                 66          sta seed+3
26  step2                       67          rts
27          jsr copy            68  up
28          ldx #17             69          asl tmp+0
29  loop2                       70          rol tmp+1
30          jsr down            71          rol tmp+2
31          dex                 72          rol tmp+3
32          bne loop2           73          rts
33          jsr merge           74  down
34          rts                 75          lsr tmp+3
35  step3                       76          ror tmp+2
36          jsr copy            77          ror tmp+1
37          ldx #5              78          ror tmp+0
38  loop3                       79          rts
39          jsr up
40          dex
41          bne loop3
    (xorshift32.s)
```

which is not as indecipherable as it may seem. The entry point is line 12, x32, which calls three subroutines before exiting. Each of these subroutines implements one line of the xorshift algorithm for a 32-bit seed stored in seed (lines 3–6). Subroutine step1 implements s ^= (s<<13), step2 implements s ^= (s>>17), and step3 implements s ^= (s<<5) to complete the algorithm. When x32 exits the next seed in the sequence is in seed.

Each subroutine calls `copy` to move the current seed value into a temporary location (`tmp`) and then performs the necessary 32-bit shifts up or down. A single bit shift up is accomplished by calling `up` and similarly a shift down calls `down`. The `asl` instruction (line 69) shifts the first byte of the copied seed up one bit position. The bit pushed off the end is stored in the 6502 carry. Then, three `rol` instructions finish the shift by shifting each byte up one bit, moving the carry bit in on the right and storing the bit lost on the left in the carry. A 32-bit shift down works in much the same way but it moves to the right starting with the highest byte (line 75).

In order to run this code we need to assemble it. The code targets memory location `0x300` and when fully assembled is,

```
300:01 00 00 00 00 00 00 00 20 12 03 20 21 03 20 30
310:03 60 20 3f 03 a2 0d 20 7d 03 ca d0 fa 20 58 03
320:60 20 3f 03 a2 11 20 8a 03 ca d0 fa 20 58 03 60
330:20 3f 03 a2 05 20 7d 03 ca d0 fa 20 58 03 60 ad
340:00 03 8d 04 03 ad 01 03 8d 05 03 ad 02 03 8d 06
350:03 ad 03 03 8d 07 03 60 ad 04 03 4d 00 03 8d 00
360:03 ad 05 03 4d 01 03 8d 01 03 ad 06 03 4d 02 03
370:8d 02 03 ad 07 03 4d 03 03 8d 03 03 60 0e 04 03
380:2e 05 03 2e 06 03 2e 07 03 60 4e 07 03 6e 06 03
390:6e 05 03 6e 04 03 60
```

which we can load into a binary file on the disk image used by the AppleWin emulator. Call it `X32.BIN`. Finally, we can use this code to estimate π with a simple BASIC program,

```
10 HOME:D$=CHR$(4):PRINT D$;"PR#3"
15 PRINT D$;"BLOAD X32.BIN"
20 PRINT "ESTIMATE PI USING XORSHIFT32": PRINT
25 C=0: N=400000
27 FOR I=1 TO N
30 CALL 776: Z=PEEK(768)+256*PEEK(769)+
   65536*PEEK(770)+16777216*PEEK(771):
   X=Z*2.32830644E-10
35 CALL 776: Z=PEEK(768)+256*PEEK(769)+
   65536*PEEK(770)+16777216*PEEK(771):
   Y=Z*2.32830644E-10
40 IF SQR(X*X+Y*Y) <= 1.0 THEN C=C+1
45 VTAB 3:PRINT "PI IS APPROXIMATELY =      ":
   VTAB 3:HTAB 23: PRINT 4.0*C/I
50 NEXT I
```

where we call the generator twice, in lines 30 and 35, to create a 32-bit value in z. We then multiply by $1/2^{32}$ to get the x and y coordinates of the point we are checking. Running this program with the seed value of 1 gives us an estimate of $\pi \approx 3.14691$ after slightly more than 400,000 iterations.

The generators of this section are extremely fast and pass tough tests of randomness. They also have very long periods. But, there is a family of generators with periods which are so long they put even the Mersenne Twister to shame. We will consider these next.

2.5 Complimentary Multiply-with-Carry Generators

The Complimentary Multiply-With-Carry (CMWC) family of generators by Marsaglia [16] are straightforward to implement and provide very long periods, well beyond that of the Mersenne Twister. Note that Marsaglia consistently referred to the generators as "Complimentary" as opposed to "Complementary" so we will use his naming convention here though we concede that "complementary" reflects the actual operation.

The code for these generators is only slightly more complex than the xorshift family though there is a state that must be initialized prior to calling the generator. The implementation given in [16] is,

```
1   #define R 4096
2   #define A 18782ULL
3   uint32_t Q[R];
4   uint32_t c=362436; // c < 809430660
5
6   uint32_t CMWC(void) {
7       static uint32_t i=R-1;
8       uint64_t t;
9       uint32_t x, m=0xFFFFFFFE;
10
11      i = (i+1) & (R-1);
12      t = A*Q[i] + c;
13      c = t >> 32;
14      x = t+c;
15      if (x < c) { x++; c++; }
16      Q[i] = m-x;
17      return Q[i];
18  }
```
(cmwc.c)

that has a state, Q, which must be initialized to seed the generator. We can use xorshift to quickly initialize the state like so,

```
1   void CMWC_init(void) {
2       int i;
3       for(i=0; i<R; i++)
4           Q[i] = xorshift();
5   }
```
(cmwc.c)

and, as usual, we can convert the output to a floating-point number by dividing by 2^{32}. We do not show the code for xorshift, it is the same as xorshift32 above where we set the seed of xorshift to determine the values written to Q. Of course, Q could be initialized with any set of 4096 unsigned 32-bit integers. The values used in the code above are those given by Marsaglia in [16] and posted online in the comp.lang.c newsgroup in 2003 [17]. Any c less than 809430660 could be used.

The generator is parameterized by R and A. Note carefully that A is explicitly defined to be of type uint64_t. The period of the generator is ab^r where a is A, r is R and $b = 2^{32} - 1$. Other a and r pairs exist for generators with different periods, see [16].

The period of this generator is $ab^r = a(2^{32} - 1)^r = 18782(2^{32} - 1)^{4096}$ which is approximately 2^{131086} or a decimal number of 39461 digits. The period of the Mersenne Twister has only 6002 digits meaning the CMWC generator's period is some 10^{33459} times longer. There is no meaningful way to put this number in human terms.

A linear congruential generator iterates an equation of the form,

$$x_n = ax_{n-1} + c \bmod m$$

for a given multiplier (a), carry (c), and modulus (m), as we saw earlier in this chapter. For this generator these are constants. If we let c change, we have a *multiply-with-carry (MWC)* generator, which we can write as iterating,

$$x_n = ax_{n-1} + c_{n-1} \bmod b, \ c_n = \lfloor (ax_{n-1} + c_{n-1})/b \rfloor$$

where there is now a subscript on c and we have changed m to b, the base of an MWC generator. As explained in [16], the multiplier (a) is chosen so that $ab - 1$ is a prime number. This will give the full period possible. The seed value is now a pair, (x_0, c_0), which is chosen at random.

One can go further and introduce a lag, denoted by r, which pulls not the previous value in the sequence but the $n - r$-th previous value. In that case, we want $ab^r - 1$ to be a prime and the generator will require a set of r random seeds along with c_0. The seed values of this generator cannot all be zero. The iteration becomes,

$$x_n = ax_{n-r} + c_{n-1} \bmod b, \ c_n = \lfloor (ax_{n-r} + c_{n-1})/b \rfloor$$

for $n \geq r$ and initial values $x_0, \ldots, x_{r-1}$ and c_0. This is a lag-r MWC generator.

There are certain problems in the equidistribution of the bits of an MWC that are addressed by using not x but the complement of x which is simply $(b - 1) - x$. For example, if $b = 2^8 = 256$ and $x = A9_{16} = 10101001_2$ the expression $(b-1)-x = 56_{16} = 01010110_2$ which is the complement of x (i.e., flip the value of every bit).

So, if we update the iteration to be,

$$x_n = (b - 1) - [ax_{n-r} + c_{n-1} \bmod b], \ c_n = \lfloor (ax_{n-r} + c_{n-1})/b \rfloor$$

we will have arrived at the *complimentary multiply-with-carry* generator (again using Marsaglia's name for it).

If we use the code above to generate a file of random bytes (assuming good equidistribution in all bits) we get a Monty Hall probability of winning to be 66.6650% for 20 million simulations and a random π value, using four bytes at a time, of 3.1418440. These are good indicators that the generator code is working properly.

What if we initialize poorly? The values in the previous run used xorshift to initialize the CMWC generator. If we simply initialize all seed values to 1 we get a win probability of 65.5364% and a π value of 3.1473600, both of which are not particularly good. Initializing with 1, 2, . . . , 4096 leads to 66.6390% and 3.1419040 which is better but not as good as using xorshift. There is more to explore here, but it is evident that the CMWC generator should be initialized well. The generator is used primarily for its extremely large period, of course, otherwise one could simply use one of the xorshift variants of the previous section.

Now that we have an idea of how the generator works, let's look again at the heart of the code to see what it is doing,

```
1   uint32_t CMWC(void) {
2       static uint32_t i=R-1;
3       uint64_t t;
4       uint32_t x, m=0xFFFFFFFE;
5
6       i = (i+1) & (R-1);
7       t = A*Q[i] + c;
8       c = t >> 32;
9       x = t+c;
10      if (x < c) { x++; c++; }
11      Q[i] = m-x;
12      return Q[i];
13  }
```
(cmwc.c)

Line 2 initializes i to be 4095. This is the index into our set of lag values in Q which are also the initial seeds of the generator. We use a 64-bit temporary, t, and a constant to represent $(b-1)$, m. Line 6 bumps the index into Q mod 4095, i.e., cycle through Q wrapping around at the end. Line 7 is the first part of the calculation. Since both A and Q[i] are 32-bit values their product is a 64-bit value which we put in t. Line 8 and 9 set the new carry, c, to the top half of t and sets x to the bottom half of t plus the carry. We get the bottom half because x is an unsigned 32-bit value. Line 10 checks for the edge case of $x < c$ and corrects for it. Line 11 replaces the old lag value in Q with the complement of the newly computed one which we return in line 12.

We can see from the code that CMWC will execute quickly, not as quickly as xorshift and its variants, but with the same sort of runtime as we saw with LCG generators like MINSTD. Of course, the 4096 word lag register in Q might be a consideration in a limited hardware environment, but that seems an unlikely setting for this generator.

2.6 Counter-Based Generators

In 2011 Salmon et al. introduced *counter-based pseudorandom number generators* [18]. All of the pseudorandom generators we have encountered in this chapter follow the same pattern:

1. Define an initial state, S_0.
2. Apply a transformation, f, to the current state to get the next state: $S_{i+1} = f(S_i)$. The new state, S_{i+1}, is the source of the pseudorandom value returned.
3. Repeat from Step 2 for each pseudorandom number needed.

If we want the 10 millionth value in the pseudorandom sequence generated above we need to apply f 10 million times.

The Salmon paper introduced a new approach. It asked the question: are there functions that create a one-to-one (bijective) mapping from the integers to another set of integers where the second set will pass, as a sequence, strict randomness tests?

Call such a function g. If such functions exist then the 10 millionth element of the sequence is simply $g(10, 000, 000)$ without requiring evaluation of *any* of the preceding values. So, if we have g we can generate a pseudorandom sequence by a simple loop calculating $g(n)$ for $n = 0, 1, 2, \ldots$.

At first glance this isn't really a benefit because all the generators above do essentially the same thing. We will see in Chap. 5 that this property of counter-based generators makes parallel application a breeze. Here we will simply introduce them and show some examples using them to make the library useful to the reader. In Chap. 4 we will see that counter-based generators pass strict randomness tests.

The central insight of [18] is that for a long time cryptographers have been building functions that are almost what a counter-based generator should be. This enabled the introduction of three new generators, two of which are based on cryptographic hashing functions. The generators are: ARS (based on AES encryption), Threefry (based on Threefish encryption), and Philox (non-cryptographic). We will investigate Threefry and Philox in this section. The ARS algorithm is best implemented on systems which support the AES-NI instructions (most x86 architectures). Note that neither Threefry nor Philox are cryptographically secure.

The Threefry and Philox generators are available for non-commercial and commercial use from

http://www.deshawresearch.com/resources_random123.html

via the *Random123* package. The implementation is for both C and C++ and consists of a series of header files only as the algorithms are implemented in macros.

Install the Random123 package with,

```
$ tar xzf Random123-1.09.tar.gz
```

where we are using the latest version available at the time of this writing. Other versions will work similarly. If desired, one may `cd` to the `examples` directory and `make` to test the library. This is highly recommended since the algorithms are implemented as macros and are very dependent upon the particular compiler switches used.

In the `examples` directory is `simple.c`,

```
1   #include <Random123/threefry.h>
2   #include <stdio.h>
3   #include <stdlib.h>
4
5   void main(void){
6       int i;
7       threefry2x64_ctr_t ctr = {{0,0}};
8       threefry2x64_key_t key = {{0xdeadbeef, 0xbadcafe}};
9       printf( "The first few randoms with key %llx %llx\n",
10          (uint64_t)key.v[0], (uint64_t)key.v[1] );
11      for(i=0; i<10; ++i) {
12          ctr.v[0] = i;
13          threefry2x64_ctr_t rand = threefry2x64(ctr, key);
14          printf("ctr: %llx %llx threefry2x64(ctr, key):
                    %llx %llx\n",
15              (uint64_t)ctr.v[0], (uint64_t)ctr.v[1],
16              (uint64_t)rand.v[0], (uint64_t)rand.v[1]);
17      }
18  }
```
(random123_simple.c)

where we have taken the liberty of slightly altering the code to explicitly call out that we are using 64-bit arithmetic.

This example uses the Threefry generator. This generator is parameterized by a key (line 8) of two 64-bit values. These serve as seeds for the generator. The counter portion is defined in line 7 though in actual use it is set as in line 12 where the first element is the index into the sequence.

The return type for this version has `threefry2x64` as a prefix. These generators return multiple values per call and operate in a 32-bit or 64-bit space. The version above returns two 64-bit values per call. The call itself is in line 13,

```
threefry2x64_ctr_t rand = threefry2x64(ctr, key)
```

where `rand` is of the same type as the counter with the two return values in `v[0]` and `v[1]`. If we run the code we get,

```
The first few randoms with key deadbeef badcafe
ctr: 0 0 threefry2x64(ctr, key): dc1e842f4112bf11
     46a388d784b2f52d
ctr: 1 0 threefry2x64(ctr, key): 60a646481acb081a
     dd49fc6023718fc2
ctr: 2 0 threefry2x64(ctr, key): eee8a29306f467ca
     1b88e50b14435b8b
ctr: 3 0 threefry2x64(ctr, key): b544060c8cb1a658
     29ec365ecdab0402
ctr: 4 0 threefry2x64(ctr, key): 44495a5bab8b1677
     392368c679c80bff
ctr: 5 0 threefry2x64(ctr, key): 319a58f11e1dc63c
     c611918ed445030f
```

```
ctr: 6 0 threefry2x64(ctr, key): 673ae0682f10b0e5
     196dde621618b964
ctr: 7 0 threefry2x64(ctr, key): a8f221275729a06f
     37604d5a2cc6adf3
ctr: 8 0 threefry2x64(ctr, key): 7e92af6203517c5b
     12d8424b33b99ed1
ctr: 9 0 threefry2x64(ctr, key): 7b4caa0c35ef583c
     dc06c31a464a68
```

The `ctr` part on the left is the index running from 0 through 9. The random output values are the right two columns in hexadecimal. This generator was parameterized by two whimsical 64-bit values, `0xdeadbeef` and `0xbadcafe`. A hallmark of counter-based generators is their sensitivity to the key with which they are parameterized. For example, if we use a slightly different key changing `0xbadcafe` to `0xbadcaff` we get,

```
ctr: 0 0 threefry2x64(ctr, key): c082867e94fee9ec
     9849a998cadf5b06
ctr: 1 0 threefry2x64(ctr, key): a1803ee31a12de15
     e928262749c9cc69
ctr: 2 0 threefry2x64(ctr, key): 18a753b405a540a9
     34d3006b432d3348
ctr: 3 0 threefry2x64(ctr, key): 2262324c33d26977
     672139fa92a1ae60
ctr: 4 0 threefry2x64(ctr, key): 45b07a9a07d69555
     be789ddbd6b57a96
ctr: 5 0 threefry2x64(ctr, key): 33f696f0ed0ba980
     838a053b1f249b5e
ctr: 6 0 threefry2x64(ctr, key): 60822997406b8439
     37d155dc053cf0f
ctr: 7 0 threefry2x64(ctr, key): 26fa4eff8afa363b
     e40df29d3e9bfdc6
ctr: 8 0 threefry2x64(ctr, key): ba63f5d406cd5832
     ee660d230ae36082
ctr: 9 0 threefry2x64(ctr, key): 8b4c6246d7b50e03
     93bdc0d434f537c1
```

which is a completely different sequence.

The Philox counter-based generator is also part of the Random123 package. It is used in a similar manner to the Threefry algorithm as this simple modification of the Threefry example above shows,

```
1  #include <Random123/philox.h>
2  #include <stdio.h>
3  #include <stdlib.h>
4
5  void main(void){
6      int i;
7      philox4x32_ctr_t  ctr = {{0,0}};
8      philox4x32_key_t  key = {{0xdeadbeef, 0xbadcafe}};
9      printf( "The first few randoms with key %lx %lx\n",
```

```
10          (uint32_t)key.v[0], (uint32_t)key.v[1]);
11       for(i=0; i<10; ++i){
12          ctr.v[0] = i;
13          philox4x32_ctr_t rand = philox4x32(ctr, key);
14          printf("ctr: %lx %lx philox4x32(ctr, key):
                  %lx %lx %lx %lx\n",
15              (uint32_t)ctr.v[0], (uint32_t)ctr.v[1],
16              (uint32_t)rand.v[0], (uint32_t)rand.v[1],
17              (uint32_t)rand.v[2], (uint32_t)rand.v[3]);
18       }
19  }
```
(random123_philox_simple.c)

In this example we include the Philox header file (line 1) and instead of returning two 64-bit values per call we are returning four 32-bit values via the call to philox4x32 (line 13). The method for setting the index and the key remain the same. A run of this program produces,

```
The first few randoms with key deadbeef badcafe
ctr: 0 0 philox4x32(ctr, key): d1e6c4f1 7c8da2e5 7569ee51
     e0e41167
ctr: 1 0 philox4x32(ctr, key): b5e7e14c 493d8f55 9584d523
     1e2d2b85
ctr: 2 0 philox4x32(ctr, key): e9b689e7 3ed0ffe5 78c28e33
     5aea3f95
ctr: 3 0 philox4x32(ctr, key): 195f3d7e da8e9336 825891d5
     50cc1447
ctr: 4 0 philox4x32(ctr, key): a3e0597d 7215a37e 6c56facd
     5d0fbfe1
ctr: 5 0 philox4x32(ctr, key): bab534e9 55b3900f 513cf0ac
     693558f1
ctr: 6 0 philox4x32(ctr, key): 9478bfa6 d2d02982 680a09a
     4ee1e7ab
ctr: 7 0 philox4x32(ctr, key): 9c03ccbe a9cc27e9 cfd2ae6d
     92ea8ad6
ctr: 8 0 philox4x32(ctr, key): a86f8d42 3a1a40f 2da41e13
     1c7174e1
ctr: 9 0 philox4x32(ctr, key): 1493ef80 92d1eb96 e20f1974
     51a00074
```

showing the four samples generated by each call to philox4x32.

The example programs are straightforward enough but some users might find it simpler still to have an interface that mimics the other generators of this chapter. For those users we can define two simple wrapper functions that only call the core routines when a new value is needed. For example, we can configure the Philox generator via,

```
 1  philox4x32_key_t philox32key = {{0,0}};
 2
 3  void philox32_init(uint32_t seed) {
 4      philox32key.v[0] = seed;
 5  }
 6
 7  uint32_t philox32(void) {
 8      static c=4,i=0;
 9      static philox4x32_ctr_t r;
10      uint32_t ans;
11      philox4x32_ctr_t ctr={{0,0}};
12
13      if (c==4) {
14          ctr.v[0] = i;
15          r = philox4x32(ctr, philox32key);
16          c=1;
17          i++;
18          ans = r.v[0];
19      } else {
20          ans = r.v[c];
21          c++;
22      }
23      return ans;
24  }
    (cbrng.c)
```

where `philox32_init` sets the seed and successive calls to `philox32` return the next 32-bit value in the sequence using a call the underlying `philox4x32` function on every fourth call. We can do the same thing for Threefry,

```
 1  threefry2x64_key_t threefry64key = {{0,0}};
 2
 3  void threefry64_init(uint64_t seed) {
 4      threefry64key.v[0] = seed;
 5  }
 6
 7  uint64_t threefry64(void) {
 8      static c=2,i=0;
 9      static threefry2x64_ctr_t r;
10      uint64_t ans;
11      threefry2x64_ctr_t ctr={{0,0}};
12
13      if (c==2) {
14          ctr.v[0] = i;
15          r = threefry2x64(ctr, threefry64key);
16          c=1;
17          i++;
18          ans = r.v[0];
```

```
19        } else {
20              ans = r.v[c];
21              c++;
22        }
23        return ans;
24    }
```
(cbrng.c)

What is the period of these generators? The bijective mapping from integers to outputs implies that for a 32-bit version the period is at least 2^{32} and for a 64-bit version the period is at least 2^{64}. However, since each generator outputs more than one value per input index the actual periods are $n2^p$ for n outputs per call and p bit outputs. Therefore, the Threefry generator above has a period of $2(2^{64}) = 2^{65}$ and the Philox generator has a period of $4(2^{32}) = 2^{34}$. This is for each parameterization by a seed value.

Let's now turn our attention to ways in which generators could be combined to increase their period.

2.7 Combined Generators

If one generator has a period, A, and another, B, it is natural to wonder if there is a way to combine them to create a good generator with a period that is some function of both. Not surprisingly, this has been done. In this section we will examine two approaches, that of L'Ecuyer [19] where two or more linear congruential generators are combined to produce a single generator that has a period much longer than any one generator alone and the KISS11 generator of Marsaglia [20] that combines multiple generators of completely different types.

In [19] the combination depends upon two lemmas proved by L'Ecuyer. The first is that if W_1 is a discrete uniform random variable between 1 and $d - 1$ for some d, which means the probability of any particular value is always $1/d$, then for any other set of discrete random variables, $W_2, W_3, \ldots$, the sum including W_1 is also uniformly distributed. The second is that for a set of generators, $s_{j,i}$, $i = 1, 2, \ldots$, with seed values in vector form as $s_{0,i} = \{s_{0,1}, s_{0,2}, s_{0,3}, \ldots\}$ and periods $p_1, p_2, p_3, \ldots$, the overall period is the least common multiple of the individual periods.

We already saw in Sect. 2.2 that we define an LCG as,

$$x_{n+1} = (ax_n + c) \bmod m$$

for suitable constants a, c, and m. Additionally, we saw that if we set $c = 0$ we have a special form known as a multiplicative LCG or MLCG. It is two or more of these that are combined in [19].

Specifically, the combined generator has the form,

$$Z_i = \left(\sum_{j=1}^{l} (-1)^{j-1} s_{j,i} \right) \bmod (m_1 - 1)$$

where Z_i is the i-th output value, j indexes over the l specific generators being combined ($s_{j,i}$) and all is taken modulo $m_1 - 1$. The sign change keeps the output value in a reasonable range instead of consistently adding. Each $s_{j,i}$ is an MLCG where j indexes the generator and i is the i-th output of that generator. Once we have a Z_i we can form a floating-point representation by dividing by m_1 unless $Z_i = 0$ in which case we return $(m_1 - 1)/m_1$.

A naive implementation to combine a set of generators, in C and using Schrage decomposition, is,

```
 1  #define L 5
 2  int32_t m[] = {2147483647,2147483647,2147483647,2147483647,
    2147483647};
 3  int32_t a[] = {39373, 742938285, 950706376, 16807,
    630360016};
 4  int32_t q[] = {54542, 2, 2, 127773, 3};
 5  int32_t r[] = {1481, 661607077, 246070895, 2836, 256403599};
 6  int32_t s[] = {1802, 6800, 6502, 8080, 4004};
 7
 8  int32_t combined_rng() {
 9      int32_t Z=0, j, k;
10
11      for(j=0; j<L; j++) {
12          k = s[j] / q[j];
13          s[j] = a[j] * (s[j] - k*q[j]) - k*r[j];
14          if (s[j] < 0) s[j] += m[j];
15          if ((j%2)==1)
16              Z = (Z - m[0] + 1) + s[j];
17          else
18              Z = Z - s[j];
19          if (Z<1) Z += (m[0]-1);
20      }
21      return Z;
22  }
```
(combined_general.c)

where we are combining the output of five MLCG generators, each with the same m value but different a values. These values are taken from Table I of [19]. The q and r vectors are for the Schrage decomposition for each of the MLCG and the five seeds are in s along with suitably computer-oriented default values.

The output is in Z, in this case a *signed* 32-bit integer. The loop over j simply calculates the new seed value for each of the five generators. Lines 15 through 18 add the newly updated seeds to the output, Z, alternating the sign for each generator. Line 19 accounts for the $Z = 0$ case. The final value is returned in line 21.

A simple driver program will exercise the combined generator,

```
1   void main() {
2       FILE *g;
3       uint32_t i,b;
4
5       g = fopen("random_combined.dat", "w");
6       for(i=0; i<10000000; i++) {
7           b = (uint32_t)combined_rng();
8           fwrite((void *)&b, sizeof(uint8_t), 3, g);
9       }
10      fclose(g);
11  }
```
(combined_general.c)

where we are only dumping the lower three bytes of each Z to disk (little-endian assumed). This is because the output is a positive signed number so the top-most bit will not be set and we want the bytes in the disk file to be as random as possible. If we run this program and look at the output of our naive test programs from this chapter we get the expected results.

Naturally, the code above is not optimized in any way. In [19] code is given for a two-MLCG combined generator with an overall period of about $2.30584 \times 10^{18} \approx 2^{61}$ which is many orders of magnitude longer than the individual periods of about $2^{31} - 1$. The code in C is,

```
1   int32_t s1 = 12345;
2   int32_t s2 = 53211;
3
4   int32_t combined_rng2() {
5       int32_t Z,k;
6       k = s1 / 53668;
7       s1 = 40014 * (s1 - k * 53668) - k * 12211;
8       if (s1 < 0) s1 += 2147483563;
9       k = s2 / 52774;
10      s2 = 40692 * (s2 - k * 52774) - k * 3791;
11      if (s2 < 0) s2 += 2147483399;
12      Z = s1-s2;
13      if (Z < 1) Z += 2147483562;
14      return Z;
15  }
```
(combined_general.c)

where the magic numbers are simply the a and m of the two selected MLCG generators and the associated $q = m/a$ and $r = m$ mod a of the Schrage decomposition. The selected parameters are, which correspond to two MLCG

a	m
40014	2147483563
40692	2147483399

generators with good performance characteristics.

In January of 2011 Marsaglia posted the code of what has become known as the KISS11 generator to the `sci.crypt` newsgroup. This was only one month before his death. While some have tried to use this generator for cryptographic purposes [21] demonstrates that it is not suitable for such use. Nonetheless, the generator has good properties and a period, according to Marsaglia, which is "almost certainly greater than" $10^{40000000}$!

The generator combines three separate generators: a linear congruential generator, an xorshift generator, and a multiply-with-carry generator. The 32-bit code, altered slightly for clarity from Marsaglia's original version, is,

```
1   uint32_t Q[4194304];
2   uint32_t carry=0;
3   uint32_t cng=123456789;
4   uint32_t xs=362436069;
5
6   uint32_t xorshift32(void) {
7        xs^=(xs<<13);
8        xs^=(xs>>17);
9        xs^=(xs<<5);
10       return xs;
11  }
12  uint32_t lcg32(void) {
13       cng = 69069*cng + 13579;
14       return cng;
15  }
16  uint32_t mwc32(void) {
17       uint32_t t,x;
18       static int j=4194303;
19       j=(j+1)&4194303;
20       x=Q[j];
21       t=(x<<28)+carry;
22       carry=(x>>4)-(t<x);
23       return (Q[j]=t-x);
24  }
25  void kiss32_init(void) {
26       int i;
27       for(i=0; i<4194304; i++) {
28            Q[i] = lcg32() + xorshift32();
```

```
29          }
30    }
31    uint32_t kiss32(void) {
32          return mwc32() + xorshift32() + lcg32();
33    }
```
(kiss32.c)

Unlike the combined LCG generators above, here we see three completely different generators combined. The combination is simply the sum, which will roll over if necessary, explaining in part why the period is (likely) so long. The main function, kiss32, returns the next unsigned 32-bit value in the sequence. Before the first call to kiss32 use kiss32_init to set up the Q register for the MWC portion. If desired, the initial seeds for the LCG and xorshift parts can be updated before calling kiss32_init by setting cng and xs respectively.

The xorshift part (lines 6–11) is the standard xorshift described above. See the exercises for suggestions in this regard. Note that the LCG portion has no explicit modulo function. This implies that $m = 2^{32}$. If we change the c value from 13579 to 1 we would be using the classic Vax random number generator. Finally, the MWC generator is a lag-r generator with $r = 4194304$ leading to a very, very long period, regardless of the contributions of the xorshift or LCG portions. Clearly, the price to pay is the overhead of storing a large buffer in RAM. For modern systems, however, this is really a minor concern.

Marsaglia also gives a 64-bit version,

```
 1    uint64_t Q[2097152];
 2    uint64_t carry=0;
 3    uint64_t cng=123456789987654321LL;
 4    uint64_t xs=362436069362436069LL;
 5
 6    uint64_t xorshift64(void) {
 7          xs^=(xs<<13);
 8          xs^=(xs>>17);
 9          xs^=(xs<<43);
10          return xs;
11    }
12    uint64_t lcg64(void) {
13          cng = 6906969069LL*cng + 13579;
14          return cng;
15    }
16    uint64_t mwc64(void) {
17          uint64_t t,x;
18          static int j=2097151;
19          j=(j+1)&2097151;
20          x=Q[j];
21          t=(x<<28)+carry;
22          carry=(x>>36)-(t<x);
23          return (Q[j]=t-x);
```

```
24  }
25  void kiss64_init(void) {
26       int i;
27       for(i=0; i<2097152; i++) {
28            Q[i] = lcg64() + xorshift64();
29       }
30  }
31  uint64_t kiss64(void) {
32       return mwc64() + xorshift64() + lcg64();
33  }
```
(kiss64.c)

which operates in exactly the same way as the 32-bit version but with constants suitable for 64-bit calculations. Note that the lag of the MWC is half that of the 32-bit version so the memory use remains the same.

This section looked at two very different implementations of combined PRNGs. How to evaluate such generators theoretically is not easy to see, especially when the algorithms involved are very different. The combined LCG generators could give some rational for the claims on their period through the two lemmas L'Ecuyer proved in [19]. The KISS generators, however, are harder to interpret. Their MWC component has a very long period, to be sure. Adding (literally) the xorshift and LCG into the mix complicates things. We saw that only the xorshift "*" (star) versions are equidistant in all bits, so it is unclear how equidistant the bits of the KISS generators actually are.

That the period of the KISS generator is very large we can see empirically. If we ignore for the time being the fact that we want to use the output of the generator for something practical we can get an idea of the period by looking to see how long it takes for three sequences that start at zero but have different individual periods to return to the state where all three are back at zero. To do this we will generate three sequences of integers starting at zero but repeating after a different number of iterations. To prevent any interplay between common factors we make the sequence length primes. This means we are asking the question: when will all the counters be back to their original all zero values?

The code is straightforward,

```
1  unsigned int A[] = {113,331,991,3001,9011};
2  unsigned int B[] = {31,97,313,919,3011};
3  unsigned int C[] = {23,61,181,419,1217};
4  unsigned int a,b,c;
5
6  unsigned int AA(unsigned int j) {  a = (a+1) % A[j];
   return a; }
7  unsigned int BB(unsigned int j) {  b = (b+1) % B[j];
   return b; }
8  unsigned int CC(unsigned int j) {  c = (c+1) % C[j];
   return c; }
```

```
 9
10  void main() {
11      unsigned int j;
12      unsigned long long i;
13
14      for(j=0; j<5; j++) {
15          i=a=b=c=0;
16          do {
17              AA(j); BB(j); CC(j);
18              i++;
19              if ((a==0) && (b==0) && (c==0))
20                  break;
21          } while (1);
22          printf("(%u,%u,%u) = %llu\n", A[j],B[j],C[j],i);
23      }
24  }
```
(combined.c)

where our three counters are in a, b, and c. For each upper limit (the modulo used
in the functions AA, BB, and CC), we ask the question: how many passes through the
loop do we need before we arrive back at the starting point of a=b=c=0? The count
is in i. The upper limits in A, B, and C are all primes, each about three times larger
than the last one.

If we run this program we see that the number of iterations, analogous to the
period of the KISS generator, increases quickly,

```
(113,31,23)       =       80569
(331,97,61)       =     1958527
(991,313,181)     =    56143123
(3001,919,419)    =  1155568061
(9011,3011,1217)  = 33019791257
```

so the claim of a period above $10^{40000000}$ for KISS does not seem to be too hard to
believe given that these sequences have periods of less than 10000 each.

Before we leave this section one general comment is in order. While there was
a time when combined generators were probably useful it could now be argued
that with the advent of good, very long period generators like xorshift1024* or
complementary MWC there is no longer really a practical reason to use them given
the uncertainties as to their properties. Then again, the hallmark of humanity is
curiosity, so practical concerns are certainly not a good enough reason to stop
research this area.

2.8   Speed Tests

In this section we will pit the generators above against each other to gain some
insight as to their relative performance. Performance here means how quickly they
generate a certain number of samples saying nothing about the quality of the

Table 2.3 Time to generate
10^9 double precision samples
(mean $\pm$ SE, 20 runs each)

Generator	Time (s)
MINSTD	9.164 ± 0.006
xorshift32	11.097 ± 0.004
xorshift128+	11.853 ± 0.028
CMWC	12.671 ± 0.003
Middle Weyl	12.849 ± 0.028
xorshift1024*	12.985 ± 0.004
Mersenne Twister	14.353 ± 0.007
KISS64	17.252 ± 0.015
Philox	18.097 ± 0.048
Threefry	32.442 ± 0.010

samples themselves. Chapter 4 will dive into testing the quality of the generators. Specifically, we will measure how long it takes for the generators to return 10^9 double precision samples. We will not store the samples, merely generate them so that the timing reflects the calls to the generator itself.

The program used is *speed_test.c* which is a slightly modified version of *ugen.c*. The source code is available on the book website (see the Preface). The generators tested are MINSTD, Mersenne Twister, xorshift32, xorshift128+, xorshift1024*, CMWC, KISS64, Middle Weyl, Philox, and Threefry. We compiled the test program using `gcc` and `-O3` optimization. Calls were timed using `clock_gettime` for fine-grained precision. All generators were seeded with the same seed value, 12345.

Table 2.3 shows the results of 20 runs, each generating 10^9 double precision samples (mean $\pm$ SE). For the system used, a 32-bit Linux system with Intel i5 processors, MINSTD is clearly the fastest generator. It is followed closely by the three xorshift variants in a group which also includes CMWC and Middle Weyl. The Mersenne Twister is slower at about 14.3 s. KISS64 and Philox come next and Threefry is last by a large amount, some $3.5\times$ slower than MINSTD. While it would certainly be possible to optimize the source code for each of these generators, along with carefully selected compiler switches, to improve performance, all the generators were placed into a single source file and compiled together to make the playing field as level as possible.

In terms of raw speed, then, and given the known claims in the sections above about the quality of the different generators, it seems clear that the tweaked xorshift variants, the complementary multiply-with-carry (CMWC), and new middle Weyl generators are all attractive options. If performance and a very long period are desired, one should consider CMWC instead of the more widely used Mersenne Twister. The new counter-based generators, Philox and Threefry, are of limited value in terms of speed. We will see their true value in Chap. 5 on parallel generators. The KISS generator's insanely long period is unlikely to be of value in any situation where a single generator is needed. There is some value in the long period in the parallel case, again, we will see this in Chap. 5.

2.9   Quasirandom Generators

Sometimes we don't want a good approximation of a random sequence. Instead, we want something that looks superficially random but is really space filling. What do we mean by "space filling"? We mean a sequence that will, as we generate more and more numbers, evenly fill the space of the unit interval, [0, 1). We insist on the unit interval because once we have that we can fill any interval, [0, a), by multiplying the sequence output by a. Further, we want to fill space in any number of orthogonal directions, i.e., for any dimensionality, d representing a d-dimensional hypercube, $[0, 1)^d$. The sequence we are looking for is called a *quasirandom* or *low-discrepancy* sequence. Fortunately, there are several ways to generate such a sequence. The one we will look at in detail in this section is the Halton sequence [22]. Another is the Sobol sequence [23].

Space filling is best illustrated via a simple example. Figure 2.7 shows plots of two collections of 500 points. The graph on the right was made by generating 1000 pseudorandom numbers, [0, 1), and plotting them in pairs. The graph on the left was made by generating 1000 quasirandom numbers in the same range and plotting them in pairs. For this simple 2D example visual inspection makes it clear that the pseudorandom generator is not filling space as nicely as the quasirandom generator is. This is the essential difference between a quasirandom generator and a pseudorandom generator. A quasirandom generator will fill space but will not pass most tests of randomness. Of course, a good pseudorandom generator will also, eventually, fill the space, but it will not do so uniformly nor without many, many samples.

Quasirandom sequences were originally intended to aid in multi-dimensional numerical integration. They are also useful for optimization problems. For exam-

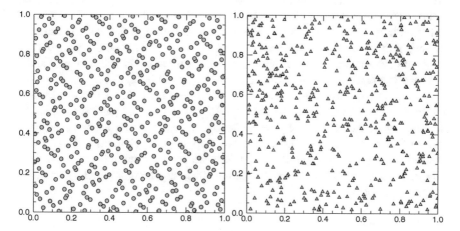

Fig. 2.7 Left: 500 points generated using the Halton quasirandom sequence. Right: 500 points generated using the NumPy default pseudorandom number generator (Mersenne Twister). By design, the quasirandom sequence does a better job of filling the 2D space with fewer points

ple, if one uses a swarm approach to locate an extremum value for a given multi-dimensional function, it might make sense to initialize the swarm with a quasirandom sequence to more fully cover the parameter space to be searched which in turn might lead to faster convergence. See Chap. 7.

The Halton sequence is generated by selecting a base, b, usually a small prime, and recursively dividing the unit interval into b sub-intervals. The endpoints of these sub-intervals are then output from smallest to largest to create the space filling sequence. While this sounds like it might be a bit complex to implement, an iterative implementation is straightforward as the following Python code illustrates,

```
1  def halton(i,b):
2      f = 1.0
3      r = 0
4      while (i > 0):
5          f = f/b
6          r = r + f*(i % b)
7          i = floor(i/float(b))
8      return r
   (halton.py)
```

The first argument, i, is the i-th value in the sequence for base b. The loop in line 4 runs until i is zero where each iterations divides i by the base (line 7) keeping the integer part. The returned value, a float (r), is built by adding the next fraction of the next smallest sub-interval of the unit interval divided into b sections. This is lines 5 and 6.

The function above gives a one dimensional output. If we want to fill d-dimensional space, we need d such sequences. To generate them, we select a set of coprime bases, which we get if we select all prime bases, and generate the sequence for each one. We then form the successive outputs into d tuples to represent the points of the d-dimensional hypercube we are trying to fill. For example, we might call the function 500 times to get sequence values 1 through 500 with a base of 2. We might then call it 500 times with a base of 3. If we form 2D points using the i-th value of each sequence, i.e., $(x, y) \leftarrow (m_i, n_i)$, with m the base 2 sequence and n the base 3 sequence, we would get the points on the left in Fig. 2.7.

One issue with the Halton sequence is that larger prime bases become linearly correlated. For example, Fig. 2.8 shows a 2D plot formed by pairing the base 17 sequence with the base 31 sequence. Clearly, there are strong linear correlations among the points. These can be lessened by skipping through the Halton sequence. Instead of keeping the $i = 1, 2, 3, \ldots$ values we keep the multiples of p, a prime: $i = p, 2p, 3p, \ldots$. This is easily, though not efficiently, accomplished by wrapping the function above,

```
1  def skip_halton(i,b,p):
2      return halton(p*i,b)
```

Fig. 2.8 Strong linear
correlations of the Halton
sequences for bases 17 and 31

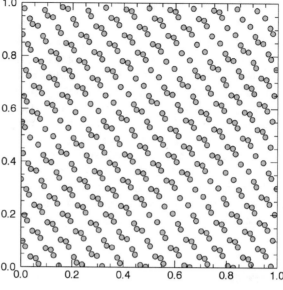

Fig. 2.9 Reduced linear
correlations of the Halton
sequences for bases 17 and 31
when using skips of
$p = 40009$

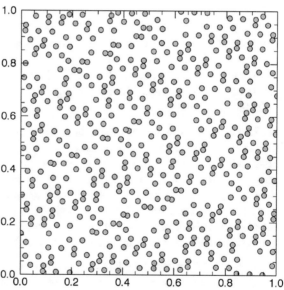

If we use the skipped Halton sequence with $p = 40009$ we get Fig. 2.9 where the
strong linear correlations have been significantly reduced. Still, they are present, so
it is advisable to use smaller primes for the base of each dimension and to limit the
overall number of dimensions when possible.

2.10   Chapter Summary

In this chapter we introduced the concept of a uniform pseudorandom number generator and developed some basic intuition and naïve test programs that we used throughout the chapter. We then described in detail generators that are of historic interest (linear congruential), in wide-spread use (Mersenne Twister), fast and good (xorshift), or of very large period (Complimentary MWC). We gave code for each, described its operation, and its strengths and weaknesses. We then looked at two representative examples of combining generators: combined LCG and KISS, to see how the period can be extended, sometimes to ridiculous levels, by combining individual generators. We next introduced counter-based generators taking a look at Threefry and Philox. We ended the chapter by taking a brief look at a standard quasirandom algorithm and explored the differences between the output of a quasirandom generator and a pseudorandom generator.

Exercises

2.1 The program to estimate π in Sect. 2.1 can be run on other files. Try running it on files of different types to see how accurate an estimate you can get. Some examples include: an audio file (raw and mp3), an executable, a large JPEG or PNG image, a bitmap image, a large Word or HTML file, a compressed .zip or .bz2 file, etc. Try to develop an intuition about the output before actually running the program on the file.

2.2 Taking a cue from the 6502 assembly version of xorshift32, write the assembly code, even if it is not possible to execute it, to implement xorshift8, an 8-bit only version with period $2^8 - 1 = 255$ for the Microchip PIC 10F200 microcontroller which is about as simple a microcontroller as one can find. The relevant documentation is on the Microchip website (http://ww1.microchip.com/downloads/en/DeviceDoc/ 40001239F.pdf) along with tools which might be helpful (search also for gpasm). Suitable constants for a full-period 8-bit xorshift generator are $(a, b, c) = (7, 5, 3)$. The XORWF, RLF and RRF instructions will be helpful here. **

2.3 Section 1.4 introduced the Iterated Function System approach to generating 2D fractals. The code there used the original version of the MINSTD generator as the "good" generator. Replace this generator with the Mersenne Twister, xorshift1024*, and CMWC (lag 4096) generators. Recall that NumPy uses the MT by default for np.random.random(). Then explore the effects on the fern and other fractals. How many iterations are necessary to see any? How far in must one zoom to notice any differences? Also recall that the fractals are of infinite extent, that zooming in to any level will still reveal the fractal structure if enough points are simulated. As a bonus, rewrite the entire program in C for speed. (** and ***)

2.4 Alter the two LCG combined generator of Sect. 2.7 to use the original version of the MINSTD generator with $a = 16807$ and the updated version using $a = 48271$. Calculate the appropriate Schrage decomposition values, q and r, for each generator, and update the code to use those values. Generate test files for the Monty Hall and random π applications and comment on the results.

2.5 The standard KISS generator, for 32-bits, uses the classic xorshift generator. Replace it with xorshift128+. Then run some simple tests to validate that it is working properly. See Chap. 4 for additional tests one could perform with this new generator.

2.6 Using the quasirandom Halton generator of Sect. 2.9 write a variant of the π estimation program and run it using the output of the generator. Compare the results to those found using some of the pseudorandom generators discussed in this chapter. Explain the results.

References

1. Lehmer, D. H. "Proceedings of the Second Symposium on Large-Scale Digital Computing Machinery." (1951).
2. Hull, Thomas E., and Alan R. Dobell. "Random number generators." SIAM review 4.3 (1962): 230–254.
3. Knuth, Donald Ervin. The art of computer programming: sorting and searching. Vol. 3. Pearson Education, 1998.
4. Marsaglia, George. "Random numbers fall mainly in the planes." Proceedings of the National Academy of Sciences 61.1 (1968): 25–28.
5. Park, Stephen K., and Keith W. Miller. "Random number generators: good ones are hard to find." Communications of the ACM 31.10 (1988): 1192–1201.
6. Marsaglia, George; Sullivan, Stephen. "Technical correspondence". Communications of the ACM. 36 (7) (1993): 105–110.
7. Schrage, Linus. "A more portable Fortran random number generator." ACM Transactions on Mathematical Software (TOMS) 5.2 (1979): 132–138.
8. Shannon, Claude E. "Programming a computer for playing chess." The London, Edinburgh, and Dublin Philosophical Magazine and Journal of Science 41.314 (1950): 256–275.
9. http://www.math.sci.hiroshima-u.ac.jp/~m-mat/MT/emt.html.
10. Matsumoto, Makoto, and Takuji Nishimura. "Mersenne twister: a 623-dimensionally equidistributed uniform pseudo-random number generator." ACM Transactions on Modeling and Computer Simulation (TOMACS) 8.1 (1998): 3–30.
11. Saito, Mutsuo, and Makoto Matsumoto. "Simd-oriented fast mersenne twister: a 128-bit pseudorandom number generator." Monte Carlo and Quasi-Monte Carlo Methods 2006. Springer, Berlin, Heidelberg, 2008. 607–622.
12. Marsaglia, George. "Xorshift rngs." Journal of Statistical Software 8.14 (2003): 1–6.
13. http://xoroshiro.di.unimi.it/ (accessed 20-Aug-2017).
14. Vigna, Sebastiano. "An experimental exploration of Marsaglia's xorshift generators, scrambled." ACM Transactions on Mathematical Software (TOMS) 42.4 (2016): 30.
15. Kneusel, Ronald T. Numbers and Computers, 2nd Ed. Springer International Publishing, 2017.
16. Marsaglia, George (2003) "Random Number Generators," Journal of Modern Applied Statistical Methods: Vol. 2 : Iss. 1, Article 2.

17. http://school.anhb.uwa.edu.au/personalpages/kwessen/shared/Marsaglia03.html (accessed 27-Aug-2017).
18. Salmon, John K., Mark A. Moraes, Ron O. Dror, and David E. Shaw. "Parallel random numbers: as easy as 1, 2, 3." In High Performance Computing, Networking, Storage and Analysis (SC), 2011 International Conference for, pp. 1–12. IEEE, 2011.
19. L'ecuyer, Pierre. "Efficient and portable combined random number generators." Communications of the ACM 31.6 (1988): 742–751.
20. Marsaglia, George (2011) "RNGs with periods exceeding 10^(40million)." Online posting 16-Jan-2011, sci.crypt.
21. Rose, Gregory G. "KISS: A bit too simple." Cryptography and Communications (2011): 1–15.
22. Halton, John H. "On the efficiency of certain quasi-random sequences of points in evaluating multi-dimensional integrals." Numerische Mathematik 2.1 (1960): 84–90.
23. Sobol, I.M. (1967), "Distribution of points in a cube and approximate evaluation of integrals". U.S.S.R Comput. Maths. Math. Phys. 7: 86–112.

Chapter 3
Generating Nonuniform Random Numbers

Abstract In this chapter we look at how to transform uniform random numbers into samples from other distributions. We only consider standard or commonly found distributions and develop a cookbook of transformations. We give code for the transformations and investigate the effects of different uniform generators on the output.

3.1 Nonuniform Random Numbers

In Chap. 2 we discussed uniformly distributed random numbers. In this chapter we will use these numbers to generate samples from other distributions. Our intention is to give a cookbook of transformations. We readily admit that we are treating the subject with less rigor than it deserves. The transformations we will discuss are for commonly used distributions and we are ignoring for now more advanced methods of generating values from arbitrary probability distributions. The guiding principle here is how can we quickly move from a uniform distribution to another distribution by using at most only a handful of draws from the uniform distribution.

The distributions we examine include the normal or Gaussian distribution, the binomial distribution, the gamma and beta distributions, the exponential distribution, and the Poisson distribution. Each of these finds frequent use in science and engineering, especially in Monte Carlo methods and in other forms of simulation. We will examine each transformation mathematically and, more importantly for us, through code.

For each of the distributions in the following sections we will use the notation U to mean a draw or sample from a uniform distribution, $[0, 1)$, using a pseudorandom number generator like any of those described in Chap. 2. If more than one uniform number is needed for a particular transform we will subscript them, $U_1, U_2, \ldots,$ where the repeated appearance of any subscripted value indicates the use of the same uniform sample. We will assume the U values are represented as C doubles meaning we will make use of any added precision available by using the conversions of Chap. 2 to directly generate an IEEE 754 double (*binary64*) value.

© Springer International Publishing AG, part of Springer Nature 2018
R. T. Kneusel, *Random Numbers and Computers*,
https://doi.org/10.1007/978-3-319-77697-2_3

To test the distributions, we will create a program to generate files of $U[0, 1)$ double precision samples for some of the generators developed in Chap. 2. These files will be used by the particular test programs for each distribution to generate histograms that will tell us how well we are doing in representing the given distributions.

The program to generate uniformly distributed doubles is straightforward. The code will make use of the MINSTD, Mersenne Twister, xorshift32, and CMWC generators of Chap. 2. Other generators will be added when necessary to illustrate a point though we will not be so pedantic as to put the code in for each case. To save space we only give the driver portion here and point the reader to the website associated with this book for the complete code. The motivated reader will be more than able to create the missing code from the snippets given in Chap. 2.

The essence of the driver program is,

```
 1  gen = atoi(argv[1]);
 2  seed = atoi(argv[2]);
 3  samples = atoi(argv[3]);
 4
 5  f = fopen(argv[4],"w");
 6  switch (gen) {
 7      case 0: seedd = seed;        break;
 8      case 1: init_genrand(seed); break;
 9      case 2: s32 = seed;          break;
10      case 3: CMWC_init(seed);     break;
11      default: break;
12  }
13  for(i=0; i < samples; i++) {
14      switch (gen) {
15          case 0: d = dminstd();        break;
16          case 1: d = genrand_real1();  break;
17          case 2: d = dxorshift32();    break;
18          case 3: d = cmwc();           break;
19          default: break;
20      }
21      fwrite((void*)&d, sizeof(double), 1, f);
22  }
23  fclose(f);
    (ugen.c)
```

where it is called with,

```
$ ugen 1 12345 2000000 double_mt.dat
```

to generate an output file, double_mt.dat, of 2,000,000 samples from the Mersenne Twister (option 1) using a seed of 12345. The operation is simple, extract command line arguments, set the seed appropriately, and then loop calling the proper uniform generator to get a double precision value written to disk. Repeat until done. The individual distribution-specific test programs will be listed in their respective sections.

3.2 Normal Distribution

Perhaps the most important nonuniform distribution is the normal or Gaussian distribution. This distribution is the familiar bell curve and is suitable for simulating many natural processes where there is a smooth distribution of values around a central mean. The normal distribution is characterized by two parameters, the mean (μ) and the standard deviation (σ). Extracting a random value from this distribution is generally written as $x \approx N(\mu, \sigma)$ where the standard zero mean unit variance (square of σ) is $N(0, 1)$. We will focus on generating $N(0, 1)$ values as we can easily map these to any μ and σ,

$$N(\mu, \sigma) = \sigma * N(0, 1) + \mu$$

There are several ways to transform uniform samples to normal samples. We will look at two of them, the Box-Muller method [1] and the polar method of Marsaglia [2]. The Box-Muller method maps two uniform samples, U_1 and U_2, to two normally distributed samples, Z_1 and Z_2, via,

$$Z_1 = \sqrt{-2 \ln U_1} \cos(2\pi U_2)$$
$$Z_2 = \sqrt{-2 \ln U_1} \sin(2\pi U_2)$$

Why does this work? If we had two independent normal coordinates, $X \approx N(0, 1)$ and $Y \approx N(0, 1)$ we could represent the 2D point they map to in polar coordinates as R, θ with $R^2 = X^2 + y^2$ and $\theta = \tan^{-1}(Y/X)$. The join probability distribution for X and Y is,

$$p(X, Y) = p(X)p(Y) = \frac{1}{\sqrt{2\pi}}e^{-\frac{X^2}{2}}\frac{1}{\sqrt{2\pi}}e^{-\frac{Y^2}{2}}$$
$$= \frac{1}{2\pi}e^{-\frac{X^2+Y^2}{2}}$$
$$= \frac{1}{2\pi}e^{-\frac{R^2}{2}}$$

which is the product of two distributions, an exponential one, $R^2 \approx \text{Exp}(1/2)$ and a uniform one, $\theta \approx U[0, 2\pi)$.

The exponential distribution is easily expressed as a transformation of a uniform sample (see Sect. 3.5). The relationship is,

$$\text{Exp}(\lambda) = \frac{-\ln U}{\lambda}$$

so that $R \approx \sqrt{-2 \ln U}$. This maps two uniform samples to the components of a 2D point using a polar representation: $(R, \theta) = (\sqrt{-2 \ln U_1}, 2\pi U_2)$. If we then map these back to Cartesian coordinates via $R\cos(2\pi\theta)$ and $R\sin(2\pi\theta)$ we will have two samples that are normally distributed. This is exactly what the Box-Muller transform does.

Code for the Box-Muller transform is,

```
 1 #define TPI 6.283185307179586231996
 2 double (*uniform)(void);
 3
 4 double norm_box(void) {
 5     static uint8_t f=0;
 6     double m,u1,u2,z1;
 7     static double z2;
 8
 9     if (f) {
10         f = 0;
11         return z2;
12     } else {
13         u1 = uniform();
14         u2 = uniform();
15         m = sqrt(-2.0*log(u1));
16         z1 = m*cos(TPI*u2);
17         z2 = m*sin(TPI*u2);
18         f = 1;
19         return z1;
20     }
21 }
```
(normal.c)

where each call to `norm_box` will return a sample from N(0, 1). Since the Box-Muller transform produces two samples at a time we use a flag (`f`) to either generate two new samples (lines 13–19) or to return the second sample in `z2` (lines 10–11). The helper function, `uniform`, must return a uniformly distributed C double in the range [0, 1). Here we define a function pointer so that we can point any function to it. This will be helpful below when we want to change the uniform generator. The generation itself is a direct mapping of the equations above returning the first sample immediately (`z1`) while stashing the second sample for the next call. This means that every other call to `norm_box` will be faster than the previous call. Note also that the Box-Muller transform is a one-to-one mapping of uniform samples to normal samples.

The sine and cosine functions of the Box-Muller transform are a bit computationally expensive. While not quite the issue it was in the past, it would be nice to remove the calls if possible. This is what the polar method attributed to Marsaglia does. We include this method though it does not use a one-to-one mapping of uniform samples to normal samples. The essential difference of this method is that the sine and cosine are calculated by sampling uniform values, taken in pairs to represent a point in 2D space, until a point within the unit circle is found. The rational is the same as for the Box-Muller transform.

If we sample uniformly from [−1, +1] until we get a point within the unit circle we can use the coordinates of that point to calculate the sine and cosine. Specifically, for a point (x, y) the sine is $y/\sqrt{s}$ and the cosine is $x/\sqrt{s}$ with $s = x^2 + y^2$. This implies that we can get two normally distributed samples with,

$$z_1 = \frac{x}{\sqrt{s}}\sqrt{-2\ln s} = x\sqrt{\frac{-2\ln s}{s}}$$

$$z_2 = \frac{y}{\sqrt{s}}\sqrt{-2\ln s} = y\sqrt{\frac{-2\ln s}{s}}$$

where we have to recall that $x^2 + y^2 < 1$ is required. This is not assured for every pair of uniform samples so the actual runtime of any implementation of this routine will be stochastic, not one-to-one as the Box-Muller transform is.

An implementation of the polar method is,

```
 1  double norm_polar(void) {
 2      static uint8_t f=0;
 3      double m,s,u1,u2,z1;
 4      static double z2;
 5
 6      if (f) {
 7          f = 0;
 8          return z2;
 9      } else {
10          do {
11              u1 = 2*uniform()-1;
12              u2 = 2*uniform()-1;
13              s = u1*u1 + u2*u2;
14          } while ((s >= 1) || (s == 0));
15          m = sqrt(-2.0*log(s)/s);
16          z1 = u1*m;
17          z2 = u2*m;
18          f = 1;
19          return z1;
20      }
21  }
```
(normal.c)

where we follow the same toggling between generating new samples and returning the second sample of a pair as we did for the Box-Muller function above. In this case, generation requires selecting pairs of uniform samples until we get a pair that is within the unit circle (lines 10–14). Here we map the uniform [0, 1) samples to [−1, +1) and try until we get $s < 1$ and not zero. Then we calculate the radial component (m) and multiply by the sine and cosine (lines 16–17). Note that we moved the square root of s into m to save a square root calculation.

The polar method was originally favored over the Box-Muller method because it was faster due to the lack of trigonometric functions. The loop will only execute once in the vast majority of cases, so while not strictly deterministic, it will still be quite fast as long as uniform is fast. However, on modern systems the advantage is minimal. If we use norm_box and norm_polar to generate two million normal samples on our test machine we get average runtimes of 175 ms versus 165 ms respectively. So, the difference is rather small but would become significant if many hundreds of millions or billions of samples were needed. For this test the uniform generator was the Mersenne Twister of Chap. 2.

We can test the normal functions by using a straightforward test program to read uniform samples from the disk files created by the ugen program of Sect. 3.1. We will use essentially the same test program for most of the distributions we examine in this chapter. The code,

```
 1  FILE *g;
 2  double next(void) {
 3      double b;
 4      fread((void *)&b, sizeof(double), 1, g);
 5      return b;
 6  }
 7
 8  void main(int argc, char *argv[]) {
 9      FILE *f;
10      int i,n;
11      double d;
12
13      uniform = &next;
14      f = fopen(argv[3], "wb");
15      g = fopen(argv[2],"rb");
16      fseek(g, 0, SEEK_END);
17      n = (int)(0.7*(ftell(g)/sizeof(double)));
18      fseek(g, 0, SEEK_SET);
19
20      if (atoi(argv[1]) == 0) {
21          for(i=0; i<n; i++) {
22              d = norm_box();
23              fwrite((void *)&d, sizeof(double), 1, f);
24          }
25      } else {
26          for(i=0; i<n; i++) {
27              d = norm_polar();
28              fwrite((void *)&d, sizeof(double), 1, f);
29          }
30      }
31      fclose(f);
32  }
```
(test_normal.c)

assigns next (line 2) to uniform (line 13) so that norm_box and norm_polar will both use it. The function simply pulls the next uniform double from the file given on the command line. This program is called with

```
$ test_normal 0|1 <uniform> <normal>
```

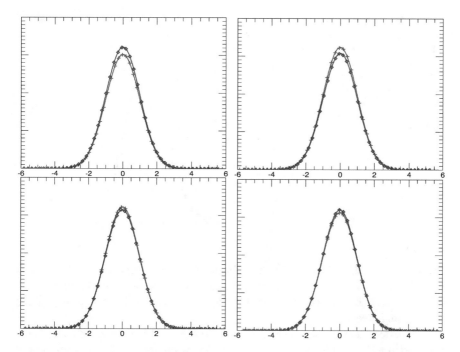

Fig. 3.1 Histograms for 70 million normally distributed samples. In each plot, the blue curve (diamonds) is generated using the polar method while the red curve (crosses) uses the Box-Muller transform. The uniform generator in each plot is, clockwise from upper left, MINSTD, Mersenne Twister, CMWC, and xorshift32 (lower left)

where a first argument of 0 will use `norm_box` and 1 will use `norm_polar`. The second argument is a file of C doubles, the output of `ugen` for a selected uniform generator. The final argument is the output file of normally distributed C doubles. Note that lines 16 through 18 determine how many output samples to generate. In this case we only output 70% of the input uniform samples. We do this because we need to account for the stochastic nature of `norm_polar`. This is not strictly required for the Box-Muller transform but we do it for both so that our analysis of the output will be based on the same number of normally distributed samples.

If we run the test program for all four uniform generators supported by `ugen` we will have four output files which should contain normally distributed C doubles. How can we test that this is so? If the output is normally distributed then a histogram of the values in the output files will follow a normal curve. Doing just this creates Fig. 3.1.

The blue (diamond) plots of Fig. 3.1 were generated with the polar method of Marsaglia. The red (crosses) plots used the Box-Muller transform. The uniform generator used was MINSTD (upper left), Mersenne Twister (upper right), xorshift32 (lower left) and CMWC (lower right). The output in all cases is clearly normally distributed and with zero mean. There are slight differences between the

Fig. 3.2 Histograms for 70 million normally distributed samples based on the RANDU uniform generator. The blue plot (diamonds) is generated with the polar method while the red plot (crosses) uses the Box-Muller method

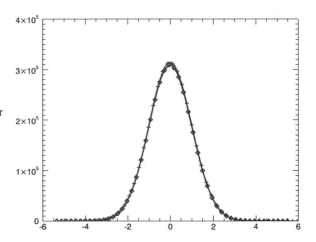

two algorithms in terms of the number of samples in the central bins but the shape is the same. If we apply the Shapiro-Wilk normality test to the source data all p-values are well above the threshold of $p = 0.05$ so we are assured of what we see, all the outputs are indeed normal.

If we extend `ugen` to output samples from the RANDU generator, which we know to be terrible at the bit level, we get Fig. 3.2. The output for RANDU is quite similar to that of the other generators. This is a good thing for old simulations using this awful generator. We ask the reader to consider why this might be so in Problem 3.2.

3.3 Binomial Distribution

If an event happens on any given trial with probability p then the probability of k events in n trials is given by the binomial distribution as,

$$P(X = k) = \binom{n}{k} p^k (1 - p)^{n-k}$$

where,

$$\binom{n}{k} = \frac{n!}{k!(n-k)!}$$

is the binomial coefficient.

If we sum the $P(X = k)$ probabilities for all possible k values, $0 \le k \le n$, we will have a discrete distribution. It is this distribution that we will learn how to draw samples from in this section.

Before we do, let's get a feel for the distribution for different p, k, and n values. To simplify things, let's fix the probability of a single event happening at $p = 0.25$ so that the distribution for fixed n is $P(X = 0) + P(X = 1) + P(X = 2) + \ldots + P(X = n)$ as a function of $k = 0, 1, 2, \ldots, n$ with each probability given by,

$$P(X = k) = \binom{n}{k}(0.25)^k(1 - 0.25)^{n-k}$$

For example, if $n = 3$ we have a discrete distribution that is,

$$P(X = 0) = \binom{3}{0}(0.25)^0(1 - 0.25)^{3-0} = 0.421875$$

$$P(X = 1) = \binom{3}{1}(0.25)^1(1 - 0.25)^{3-1} = 0.421875$$

$$P(X = 2) = \binom{3}{2}(0.25)^2(1 - 0.25)^{3-2} = 0.140625$$

$$P(X = 3) = \binom{3}{3}(0.25)^3(1 - 0.25)^{3-3} = 0.015625$$

where the total of all the probabilities is indeed 1.0, as expected. This means we can plot the distributions by varying n and p. Figure 3.3 shows different distributions for fixed $n = 30$ and various probabilities $p \in \{0.1, 0.3, 0.5, 0.7, 0.9\}$. Figure 3.4 fixes $p = 0.3333$ and varies $n \in \{10, 20, 30, 40, 50\}$. As n gets large we see that the distribution approaches a normal one, $N(\mu, \sigma)$, with $\mu = np$ and $\sigma = np(1 - p)$.

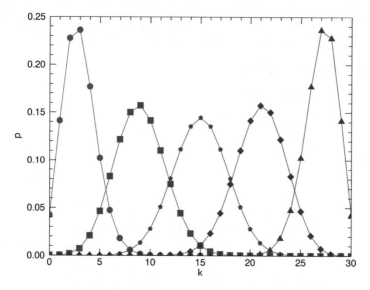

Fig. 3.3 Various binomial distributions for $n = 30$ and p of 0.1 (circle), 0.3 (square), 0.5 (star), 0.7 (diamond), 0.9 (triangle). The distributions are discrete, the lines are only present to help follow the path. Note that the distributions are symmetric in that $p = 0.1$ is the reverse of $p = 0.9$

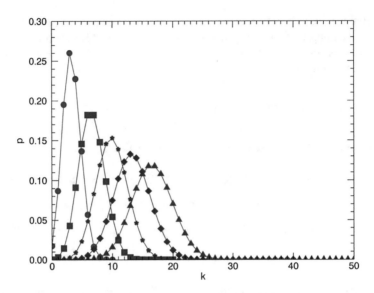

Fig. 3.4 Various binomial distributions for $p = 0.3333$ and n of 10 (circle), 20 (square), 30 (star), 40 (diamond), 50 (triangle). The distributions are discrete, the lines are only present to help follow the path

How to sample from this discrete distribution? We will consider three approaches. One immediately suggests itself from the example above with $n = 3$. Simply draw a uniform sample, $U[0, 1)$, and sum the probabilities of each k for fixed n and p until we get to the uniform sample. In code this is,

```
 1  double (*uniform)(void);
 2
 3  double fact(int n) {
 4       int i;
 5       double x=1.0;
 6       for(i=1; i<=n; i++)
 7           x *= i;
 8       return x;
 9  }
10
11  double B(int n, int k, double p) {
12       double c = fact(n)/(fact(k)*fact(n-k));
13       return c*pow(p,k)*pow(1.0-p, n-k);
14  }
15
16  int binomial_cumulative(int n, double p) {
17       int i;
18       double c=0.0, u,t;
```

```
19        u = uniform();
20        for(i=0; i<=n; i++) {
21             t = B(n,i,p);
22             if ((c <= u) && (u < (c+t)))
23                  return i;
24             c += t;
25             x++;
26        }
27        return n;
28   }
```
(binomial.c)

where, as in Sect. 3.2, we use a function pointer, uniform, to point to a function that returns uniformly distributed C doubles. Lines 3–9 implement a straightforward factorial function. Lines 11–14 calculate the probability of exactly k events happening in n trials with probability p. It is this value that we sum for each k until we reach the uniform sample. Both fact and B will be used again below. The actual sample is drawn in binomial_cumulative. This function picks a uniform sample (u) and then calculates the cumulative distribution for the given n and p until we bracket the uniform sample (line 22).

While simple to implement, and fairly obvious, the constant recomputation of the cumulative distribution on each call has serious consequences in terms of runtime performance. In order to see this, let's create a simple test program,

```
1   FILE *g;
2   double next(void) {
3        double b;
4        fread((void *)&b, sizeof(double), 1, g);
5        return b;
6   }
7
8   void main(int argc, char *argv[]) {
9        FILE *f;
10       int i,n,m,N,d;
11       double p;
12
13       m = atoi(argv[1]);
14       n = atoi(argv[2]);
15       sscanf(argv[3],"%lf",&p);
16       uniform = &next;
17       f = fopen(argv[5],"wb");
18       g = fopen(argv[4],"rb");
19       fseek(g, 0, SEEK_END);
20       N = (int)(ftell(g)/sizeof(double));
21       fseek(g, 0, SEEK_SET);
22       for(i=0; i<N; i++) {
23            switch (m) {
```

```
24              case 0:   d = binomial_cumulative(n,p);
                    break;
25              case 1:   d = binomial_recycle_coin(n,p);
                    break;
26              case 2:   d = binomial_table(n,p);
                    break;
27              default: break;
28          }
29          fwrite((void*)&d, sizeof(int), 1, f);
30      }
31      fclose(f);
32  }
```
(test_binomial.c)

where `binomial_recycle_coin` and `binomial_table` are defined below. This program will output a binary file of signed integers. Call this test program with,

```
$ test_binomial 0|1|2 <n> <p> <uniform> <binomial>
```

and use "0" to select the cumulative sample function. The `uniform` file is the output of the `ugen` program found in Sect. 3.1. Using the test program to generate 100,000 samples for $n = 100$, $p = 0.7$ on our test machine takes on average 5.5 s. Surely we can do better.

We already identified the bottleneck, it is the repeated calculation of the individual probabilities for individual k values. One way to remedy this situation is to realize that we will often want to draw many samples for a specific n and p pair. So, we can save time, at the expense of some memory, and generate the probability table once and use it until either n or p change. Doing this gives us,

```
1  double *table = (double *)NULL;
2  int ntable = 0;
3  double ptable = 0;
4
5  int binomial_table(int n, double p) {
6      int i;
7      double u;
8      if ((ntable==0) || (ptable==0) || (ntable != n) ||
              (ptable != p)) {
9          if (table != NULL)
10             free(table);
11         table = (double *)malloc((n+2)*sizeof(double));
12         ntable = n;
13         ptable = p;
14         table[0] = 0.0;
15         for(i=0; i<=n; i++)
16             table[i+1] = B(n,i,p)+table[i];
17     }
```

```
18        u = uniform();
19        for(i=0; i<n; i++)
20             if ((table[i] <= u) && (u < table[i+1]))
21                  return i;
22        return n;
23   }
```
(binomial.c)

where we are still selecting a single uniform sample (line 18) and checking the cumulative probability distribution until we bracket it (line 20) but instead of calculating the probabilities repeatedly we do it once when called with a specific n or p. The table is built in lines 8–17. We allocate memory for $k = 0, 1, 2, \ldots, n$ with a leading term of zero, hence $n + 2$. We keep the current n and p (lines 12–13) and build the table (lines 14–16). Note that we set index $i + 1$ to the probability returned by B for i and add in the last table element found, table[i].

How does the runtime performance compare with binomial_cumulative? Instead of 5.5 s we have now dramatically reduced the runtime to a mere 0.045 s! True, we need to store n doubles in memory, but that is typically not an issue and certainly a small price to pay for improving the runtime by two orders of magnitude.

Devroye gives several other algorithms for sampling from a binomial distribution, see [3, page 524] and following. One of these is particularly simple, the coin flipping algorithm. In C code it looks like this,

```
1   int binomial_naive_coin(int n, double a) {
2        double p;
3        int i,k=0;
4        p = (a <= 0.5) ? a : 1.0-a;
5        for(i=0; i<n; i++)
6             if (uniform() <= p)
7                  k++;
8        return (a <= 0.5) ? k : n-k;
9   }
```
(binomial.c)

where we've named it binomial_naive_coin for reasons that will become clear. The original algorithm in [3] is only valid for $p \leq 0.5$ but we have extended it for all p by noting that the distribution for $p \leq 0.5$ is the mirror of the distribution for $(1 - p) > 0.5$.

Specifically, line 4 checks if we need to mirror the output and sets the local p accordingly. We then flip a biased coin for each of the possible trials, n. This is line 6. If the coin flip, i.e., the $U[0, 1)$ value returned by uniform, is less than p we increment the counter in k. When we've performed all n flips we return k after adjusting a second time if the input probability a is greater than 0.5 (line 8).

This algorithm is not calculating probabilities explicitly according to theory. Rather, it is simulating a process that will produce a binomial probability distribution for the given n and p. However, this naïve algorithm has a serious problem.

Each simulation uses a separate call to uniform which means that we need to have many more uniformly distributed inputs available than we will get as binomially distributed outputs. We would like to make the mapping from uniform inputs to binomial outputs one-to-one. This is what we had with the cumulative and table versions above.

Devroye's next binomial algorithm makes this possible by noting that for each of the n trials we really only need one bit of information from the uniform number. The uniform sample can then be rescaled and reused for the next trial. The code for this version is,

```
 1  int binomial_recycle_coin(int n, double a) {
 2      double p,u;
 3      int b,i,k=0;
 4      p = (a <= 0.5) ? a : 1.0-a;
 5      u = uniform();
 6      for(i=0; i<n; i++) {
 7          b = (u > (1.0-p)) ? 1 : 0;
 8          u = (u-(1.0-p)*b) / (p*b+(1.0-p)*(1-b));
 9          k += b;
10      }
11      return (a <= 0.5) ? k : n-k;
12  }
    (binomial.c)
```

with a structure that is very similar to the naïve version. However, we only call uniform once, in line 5, and then use it to select a bit (line 7) which is the result of our trial. Line 8 rescales u so that we can use it again to get another bit for the next trial. Line 9 adds the selected bit to k which we then return in line 11 adjusting for $p > 0.5$ as before.

This function is attractive because it only uses one uniform number for each binomial output. Also, there is no trade-off with memory, no table is created. There are more operations but it is hard to beat the compactness. What about the runtime? On our test machine 100,000 samples ($n = 100$, $p = 0.7$) takes on average 0.21 s. That is 5x slower than the table method but over 20x faster than the original cumulative approach without any memory overhead.

When we tested the code to generate normally distributed values in Sect. 3.2 we saw some differences between the algorithms and the source uniform samples. The binomial table algorithm shows similar differences as seen in Fig. 3.5.

In this figure we show overlapping histograms of one million binomial samples from $n = 100$ and $p = 0.7$. What changed between each run is the uniform sample source. The uniform samples came from the MINSTD, Mersenne Twister, xorshift32, CMWC and even the awful RANDU generator. Regardless of the source, the output binomial samples are all equally good. With $n = 100$ and $p = 0.7$ we would expect a peak at bin 70, as is the case here. In Fig. 3.5 the histograms are overlapping, without bars, so that differences in each bin are represented as a spread in the vertical direction.

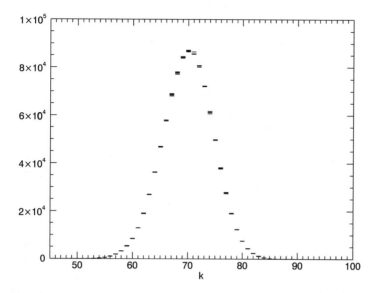

Fig. 3.5 Multiple histograms from binomial samples using the table algorithm and based on different uniform generator output. In each case, $n = 100$ and $p = 0.7$, with one million samples drawn. The uniform samples were generated with the `ugen` program described in the text using MINSTD, Mersenne Twister, xorshift32, CMWC, and the awful RANDU generator

If we consider the recycled coin flipping algorithm we still get very good agreement, see Fig. 3.6. The parameters are the same as in Fig. 3.5, only the binomial sampling algorithm has changed.

If we fix the uniform generator to be the Mersenne Twister and create five different output files using the `ugen` program with five different seed values we get Fig. 3.7 showing much the same pattern as that found with different uniform generators.

3.4 Gamma and Beta Distributions

The gamma distribution (Gamma(α, β)) is a commonly used distribution. The exponential and χ-square distributions are special cases of it. The gamma distribution depends upon two parameters, α and β, which define its shape and "rate". The distribution is foundational to Monte Carlo methods because if random samples can be drawn from a gamma distribution simple transformations will map those to other key distributions including chi-square, t, F, beta, and Dirichlet [4]. We will focus on generating beta distributed samples from gamma samples later in this Section.

There are multiple ways to parameterize the gamma distribution. We will use the following form of the probability density function,

$$y = \frac{\beta^\alpha}{\Gamma(\alpha)} x^{\alpha-1} e^{-\beta x}$$

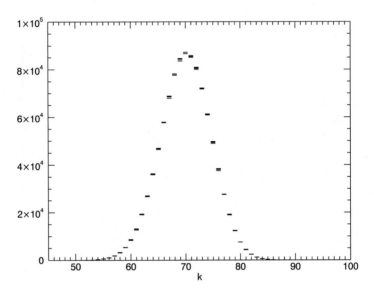

Fig. 3.6 Multiple histograms with the same sources and parameters as in Fig. 3.5. The only change here is the use of the recycled uniform algorithm instead of the table algorithm

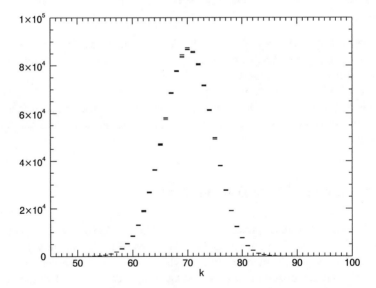

Fig. 3.7 Multiple histograms of the $n = 100$, $p = 0.7$ binomial distribution. There were one million samples drawn in each case using the recycled uniform (coin flip) algorithm as in Fig. 3.6. The uniform samples were drawn from the Mersenne Twister with a unique seed value for each histogram

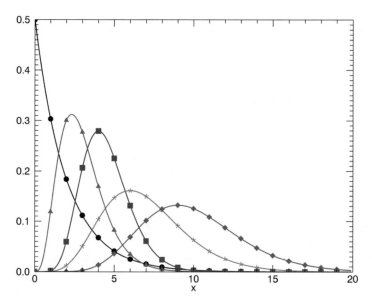

Fig. 3.8 Several different gamma distributions: Gamma(1, 0.5) (circle), Gamma(9, 2) (square), Gamma(7, 1) (star), Gamma(4.5, 1.5) (triangle), and Gamma(10, 1) (diamond)

where $\Gamma(\alpha)$ is the complete gamma function. In this parameterization, α is the *shape* and β is the *rate*. An alternative parameterization uses k for the shape and θ for the *scale*. The relationship is $\alpha = k$ and $\beta = 1/\theta$. For example, the Python NumPy library uses shape and scale so to match what NumPy generates with what our code here generates, use the reciprocal of β. To further muddy the waters, some authors use λ in place of β. Finally, if $\beta = 1$ we have the *standard gamma distribution* which is what the algorithms below sample from. A straightforward scaling by β maps the standard gamma to the rate parameter.

The closed form of the PDF makes plotting the gamma distribution straightforward. Figure 3.8 shows the distribution for several α and β values.

There are several approaches to sampling from a gamma distribution. For example, see Devroye [3], as well as Marsaglia and Tsang [4] and Ahrens and Dieter [5, 6]. We will implement two versions here starting with Algorithm GC of Ahrens and Dieter [5].

The Ahrens GC algorithm samples from the standard gamma distribution with a shape parameter greater than one. In C code the algorithm is,

```
1  double (*uniform)(void);
2  #define DPI 3.1415926535897932
3
4  double gc_gamma(double a) {
5      double b,c,s,x,f,t,u;
6      b = a - 1.0;
```

```
 7        c = a+b;
 8        s = sqrt(c);
 9        while (1) {
10            while (1) {
11                t = s*tan(DPI*(uniform()-0.5));
12                x = b + t;
13                if (x >= 0.0) break;
14            }
15            u = uniform();
16            f = exp(b*log(x/b) - t + log(1.0 + t*t/c));
17            if (u <= f) return x;
18        }
19   }
20
21   double gamma_gc(double a, double b) {
22        return (a > 1.0) ? gc_gamma(a)/b
23               : gc_gamma(a+1.0)*pow(uniform(),1.0/a)/b;
24   }
```
(gamma.c)

where we again use a function pointer, uniform, to acquire $U[0, 1)$ samples from our test program, ugen. Note, the expected number of uniform samples necessary to generate a single gamma distributed sample is a function of α (a) which approaches $\pi^{\frac{1}{2}} \approx 1.7725$ as $\alpha \to \infty$. Therefore, it is safest to generate about one tenth as many gamma variates for a given input file of uniform variates. Naturally, if uniform is a true pseudorandom generator, as would be the case if using this code in an actual application, this restriction would not be in place.

The algorithm uses acceptance-rejection sampling. According to [5] the sampling function is proportional to a Cauchy probability density whose integral is an inverse tangent. This leads to the sampling criteria,

$$x \leftarrow a - 1 + (2a - 1)^{\frac{1}{2}} \tan(\pi(u - 0.5))$$

where we generate an x (lines 10–14) from the Cauchy density. We compare this to our uniform sample u where f comes from the relationship between the gamma distribution and the Cauchy density (equation 3.3 of [5]). If u is less than f we accept x and return it (line 17).

All of the above is the function gc_gamma that samples from the standard gamma distribution for $\alpha > 1$. One way to get outputs for $\alpha \leq 1$ is seen in line 23 that first finds gamma($\alpha + 1$) and then scales by a uniform value raised to the $1/\alpha$ power (see note in [4]). Finally, to add the rate β into the output we simply scale the standard gamma by it (lines 22 and 23).

The Ahrens GC algorithm is compact but makes use of potentially expensive trigonometric and transcendental functions. If we are able to rapidly sample from a normal distribution (as in Sect. 3.2) then we can improve matters. This is exactly what the algorithm of Marsaglia and Tsang does [4].

In [4] a detailed description of a very compact method for sampling from the standard gamma distribution is given. It is also an acceptance-rejection method but makes use of samples from the normal distribution, which we can supply here through the Box-Muller transform of Sect. 3.2.

The Marsaglia-Tsang method depends on the fact that there is a relationship between normally distributed variates cubed and gamma variates. This relationship makes acceptance-rejection straightforward. The algorithm, as succinctly described in the abstract of [4] is,

Generate a normal variate x and a uniform variate U until $\ln U < 0.5x^2 + d - dv + d \ln v$ then return dv. Here, the gamma parameter is $\alpha \geq 1$, and $v = (1 + x/\sqrt{(9d)})^3$, with $d = \alpha - 1/3$.

The code in C is,

```c
1   double ms_gamma(double a) {
2       double d,c,x,v,u;
3       d = a - 1.0/3.0;
4       c = 1.0 / sqrt(9.0*d);
5       while (1) {
6           do {
7               x = norm_box();
8               v = 1.0 + c*x;
9           } while (v <= 0.0);
10          v = v*v*v;
11          u = uniform();
12          if (u < 1.0-0.0331*(x*x)*(x*x)) return d*v;
13          if (log(u) < 0.5*x*x+d*(1.0-v+log(v))) return d*v;
14      }
15  }
16
17  double gamma_ms(double a, double b) {
18      return (a > 1.0) ? ms_gamma(a)/b
19                       : ms_gamma(a+1.0)*pow(uniform(),1.0/a)/b;
20  }
```
(gamma.c)

where we have expanded the syntax of the example in [4] to enhance readability. We have also followed the form of the Ahrens code above in that we generate the standard gamma distribution sample first (ms_gamma) and then apply the scaling and shifting trick to get the two parameter sample (lines 17–20). Notice that the code above calls norm_box to find a standard normal sample. We can use the code from Sect. 3.2 here and take advantage of the fact that the code will also use the uniform function pointer. Recall that the Box-Muller transformation is one-to-one so that each uniform sample used leads to a one normal sample.

This method generates normal variates (x) and uses a specific function which ultimately cubes x to be a gamma that fits under a normal curve. The description of the search and justification for the particular function of x, namely v, is in [4]. The

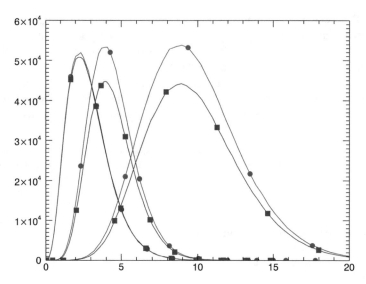

Fig. 3.9 Comparing the histograms generated by the Ahrens GC algorithm (circles) and the Marsaglia algorithm (squares). For all plots the same sequence of uniform numbers from the Mersenne Twister was used. Notice that while the shapes of the distributions are the same, the number of absolute samples in each bin is different. Left, Gamma(9, 2), middle Gamma(4.5, 1.5), and right Gamma(10, 1)

test is line 13 above, if this condition holds then the product, dv, is under the gamma curve for the given a. Line 12 is a "squeeze" [7] which is another function that is often a suitable test. This function does not have the two logarithms and is true about 92% of the time which means that the (relatively) expensive code only runs about 8% of the time (empirical testing). This improves the runtime performance, but the algorithm works without the test in line 12. Line 19 utilizes the same trick we used with the Ahrens GC algorithm above. The reference for it is the note in [4].

How do our two gamma distribution sampling methods compare to one another? In terms of output, we see in Fig. 3.9 that they produce the expected distributions but vary somewhat, depending upon α and β, as to the number of samples in each bin. Recall that these plots are actually histograms with 100 bins generated from 1,000,000 samples from the respective gamma distributions. Each distribution used the same source file of uniform samples drawn from the Mersenne Twister with the same seed value each time so the differences seen are not due to fluctuations in the uniform samples supplied to the algorithms.

We can also compare the runtime performance of each algorithm. In this case, we have two items we can track, the runtime against the clock to generate one million samples and the number of uniform samples necessary to get the one million gamma samples. As in Fig. 3.9 we again use the same source file of uniform samples in each case.

The comparison shows that on our test machine the runtime of either algorithm is really not a function of the α value used. The Ahrens algorithm takes, on average, 0.47 seconds to generate one million gamma samples while the Marsaglia-Tsang algorithm does the same in 0.21 s for a 2x performance improvement.

The number of uniform samples used varies only slightly by input parameter α,

	$\alpha = 4.5$	$\alpha = 9$	$\alpha = 10$
Ahrens GC	3,298,733	3,339,298	3,347,113
Marsaglia-Tsang	2,014,131	2,006,632	2,005,896

showing that the Marsaglia-Tsang algorithm uses less than 2/3 the number of uniform samples required by the Ahrens GC algorithm.

Let's look at the distributions we get if we sample from Gamma(4.5, 1.5) using both the Ahrens and Marsaglia-Tsang algorithms and different uniform samples from the ugen program. In each case we are generating a file of uniform C doubles, ten million, and then using these to create one million gamma distributed output samples. Finally, we will plot histograms (bins=100) to visualize the spread of outputs as a function of the uniform sample source.

Figure 3.10 shows the histograms. On the left is the output of the Ahrens algorithm, the Marsaglia-Tsang algorithm output is on the right. The uniform samples came from the MINSTD, Mersenne Twister, xorshift32, CMWC, xorshift1024*, KISS64 and RANDU generators. For the Ahrens algorithm the histograms are reasonably overlapping. The Marsaglia-Tsang algorithm, using the same uniform samples, exhibits greater diversity. In particular, two histograms are quite a bit apart from the others. These are the xorshift32 (highest) and MINSTD (second highest) histograms. Surprisingly, and encouragingly for old Monte Carlo work, the RANDU output is not significantly different from the other generators in this case, as long as a suitable seed value is chosen. For these plots, all generators used a seed of 12345, with tables of seeds built from this one value when necessary, except for RANDU which needed a different seed, 12111.

The beta distribution is another two parameter distribution used in a variety of settings because of its shape flexibility. We are including it in this section because once we are able to sample from the gamma distribution we get the beta distribution virtually for free. Specifically, the beta distribution, Beta(a, b), is easily generated from two gamma distributions,

$$\text{Beta}(a, b) = \frac{X}{X + Y}, \quad X \sim \text{Gamma}(a, \theta), \quad Y \sim \text{Gamma}(b, \theta)$$

for any θ. See Devroye [3, Theorem 4.1]. Therefore, we get samples from beta distributions with only a few lines of code,

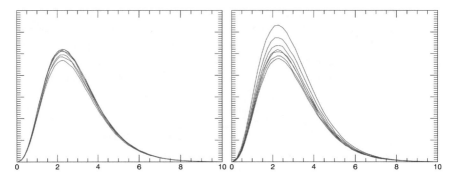

Fig. 3.10 Histograms of Gamma(4.5, 1.5) sampled using the Ahrens GC algorithm (left) and the Marsaglia-Tsang algorithm (right). The only difference between each histogram is the uniform sample source. The uniform samples were from the MINSTD, Mersenne Twister, xorshift32, CMWC, xorshift1024*, KISS64 and RANDU generators. For the Marsaglia-Tsang histograms (right) the two histograms significantly different from the others are the xorshift32 (highest) and MINSTD (second highest)

```
1  double beta_gc(double a, double b) {
2      double x = gamma_gc(a,1);
3      return x / (x + gamma_gc(b,1));
4  }
   (beta.c)
```

where we are calling the Ahrens GC gamma function above. Naturally, one could easily replace this call with a call to the Marsaglia-Tsang gamma function. Here we arbitrarily set $\theta = 1$. Notice that the x in the numerator and the denominator of line 3 must be the same, not two separate draws from Gamma(a, 1).

Figure 3.11 shows the histograms generated by one million samples from Beta(a, b) for several values of a and b. Notice that the distributions are mirror images of each other when a and b are reversed. This property can be useful, for example, simulated classification experiments were performed in [8] using Beta(θ, 1) for targets and Beta(1, θ) for nontargets. The larger θ was the more discriminative targets were from nontargets in displays interpreted by human observers.

3.5 Exponential Distribution

The exponential distribution, i.e., a distribution that follows a decaying exponential, is one of the few distributions that can be sampled directly via inverse transform sampling. The basic idea is that if we know the form of the cumulative distribution function for a distribution then we can use the inverse function to generate values

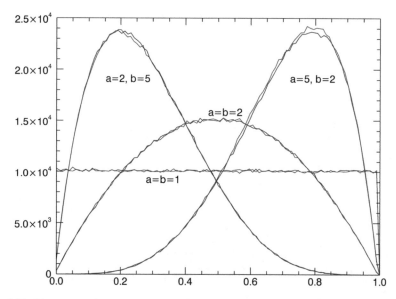

Fig. 3.11 Histograms of one million draws from different beta distributions, Beta(a, b), with a and b as shown. Distributions generated with both the Ahrens (red, if in color) and Marsaglia (blue) gamma functions are plotted, hence the overlapping lines. Notice the close agreement between the two algorithms. The uniform source was the Mersenne Twister

distributed according to the initial distribution. An example will make the process clear.

We want samples from a decaying exponential distribution that is proportional to $e^{-\lambda x}$ where λ is a scaling factor. For this distribution the cumulative distribution function (CDF) is $1 - e^{-\lambda x}$, which makes sense, as x gets larger we have more and more of the total probability density $< x$ but the full probability density takes an infinite amount of time to be reached since $e^{-\lambda x}$ is getting smaller and smaller.

If $F(x) = 1 - e^{-\lambda x}$ and $F^{-1}(u) = x$ then the inverse of $F(x)$ is,

$$F(F^{-1}(u)) = u = 1 - e^{-\lambda F^{-1}(u)}$$

$$1 - u = e^{-\lambda F^{-1}(u)}$$

$$\ln(1 - u) = -\lambda F^{-1}(u)$$

$$F^{-1}(u) = \frac{-\ln(1-u)}{\lambda}$$

where if we select $u \sim U[0, 1)$ we will map via $F^{-1}(u)$ back to the exponential distribution. Additionally, since we are selecting u at random we see that $1 - u$ and u are just complements of each other so that if $u = 0.2$ then $1 - u = 0.8$ and vice versa. This implies that we can replace $1 - u$ with u in our definition to arrive at,

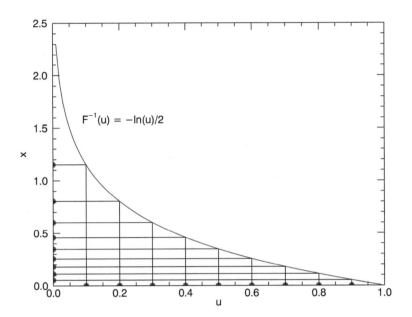

Fig. 3.12 A plot of the inverse transform of the CDF for the exponential distribution, $\lambda = 2$. Notice that uniformly spaced inputs, u, (red dots on the u-axis) map to nonuniformly spaced outputs on the x-axis. The spacing is crowded around zero which is what a decaying exponential distribution would do

$$F^{-1}(u) = \frac{-\ln u}{\lambda}$$

Figure 3.12 shows graphically what the application of $F^{-1}(u)$ is actually doing. For any u we pick on the u-axis we get an output value on the x-axis which passes through $x = F^{-1}(u)$. The mapping compresses output values to be closer to zero on the x-axis so that uniform selection on the u-axis will, in most cases, map to something closer to zero on the x-axis instead of far from it. This is exactly what we would expect an exponential distribution to do since as x gets larger the probability of it being chosen is decreasing continuously.

Since we can make use of the inverse transform in this case, the code to calculate exponentially distributed samples based on a single input uniform sample is particularly straightforward,

```
1  double exponential(double lambda) {
2      return -log(uniform()) / lambda;
3  }
```
(exponential.c)

where we simply translate $F^{-1}(u)$ into C code replacing u with a call to our uniform number generator, $U[0, 1)$.

3.6   Poisson Distribution

The final distribution of this chapter is the Poisson distribution. Like the binomial above (Sect. 3.3) the Poisson distribution is discrete. It represents the probability of a given number of events occurring in a fixed interval assuming a constant rate and independence (i.e. the events happen without influence from previous events). Examples of processes that follow the Poisson distribution include radioactive decay and photon or particle counting. For example, the number of photons incident on an x-ray detector per unit time will follow Poisson statistics. For more discussion, see Bevington [9].

The Poisson distribution is a limiting case of the binomial for a large number of trials. We can see that this is so quite nicely by looking at the binomial probability function and seeing what happens to it when the number of trials gets large ($n \rightarrow \infty$). The expected value of a binomial distribution is np for n trials each with probability p of happening. If we call this expected value λ we can write the probability for a single trial as $p = \lambda/n$. Plugging this into the binomial probability mass function gives,

$$
\begin{aligned}
P_{bin}(k) &= \binom{n}{k} p^k (1-p)^{n-k} \\
&= \binom{n}{k} \left(\frac{\lambda}{n}\right)^k (1-\frac{\lambda}{n})^{n-k} \\
&= \frac{n(n-1)(n-2)\cdots(n-k+1)}{k!} \left(\frac{\lambda^k}{n^k}\right) (1-\frac{\lambda}{n})^{n-k} \\
&= \frac{n}{n} \cdot \frac{n-1}{n} \cdots \frac{n-k+1}{n} \cdot \frac{\lambda^k}{k!} (1-\frac{\lambda}{n})^{n-k} \\
&= \frac{n}{n} \cdot \frac{n-1}{n} \cdots \frac{n-k+1}{n} \cdot \frac{\lambda^k}{k!} (1-\frac{\lambda}{n})^{n} (1-\frac{\lambda}{n})^{-k}
\end{aligned}
$$

which is useful because as $n \rightarrow \infty$ all the fraction terms tend towards 1. Additionally, the last term to the $-k$ power also tends to 1. This means that,

$$
\lim_{n \to \infty} P_{bin}(k) = \lim_{n \to \infty} \frac{\lambda^k}{k!} (1-\frac{\lambda}{n})^n = \frac{\lambda^k}{k!} e^{-\lambda} = \frac{e^{-\lambda}\lambda^k}{k!} = P_{Poisson}(k, \lambda)
$$

because $(1-\frac{\lambda}{n})^n$ as $n \rightarrow \infty$ becomes $e^{-\lambda}$. The last form on the right is the definition of the Poisson probability mass function with mean λ.

As with several of the other distributions in this chapter, there are a plethora of algorithms that we could use to draw samples from a Poisson distribution. We will implement three: the multiplication method (Knuth [10]), the inversion method (Devroye [3]) and the upward/downward search of Kemp [11].

The multiplication method is simple to implement,

```
 1   int poisson_mult(double l) {
 2        double m=exp(-l), p;
 3        int k=0;
 4        p = uniform();
 5        while (p >= m) {
 6             k++;
 7             p *= uniform();
 8        }
 9        return k;
10   }
```
(poisson.c)

where we set m to the value of the decaying exponential for the given λ (l) which is the mean value of the distribution (line 2). We then count, in k, the number of times we have to multiply an initial uniform value (line 4) by another uniform value before we are below the exponential. Line 7 will only decrease p because we are multiplying a number less than one by another number less than one.

Three difficulties immediately present themselves with this approach. First, we are using many uniform samples to create a single sample from the Poisson distribution. Second, this algorithm is linear in time with respect to λ. Third, as λ increases round-off error from the calculation of $e^{-\lambda}$ and from the repeated multiplication in line 7 causes values to appear in the output with the wrong frequency. To see this, look at Fig. 3.13.

In this figure we see histograms for $\lambda \in \{4, 8, 12, 24\}$. While the histograms are correct for $\lambda = 4$ and $\lambda = 8$ there are spurious counts of $k = 0$ for the larger λ values. These are due to floating-point rounding errors and imply that this algorithm is only useful for small λ values (< 10).

The multiplication method is not satisfying. Let's see if we can do better. Devroye [3] gives another simple algorithm based on the ratio of the probabilities for measuring k and $k + 1$ events. Specifically,

$$\frac{P(k = i + 1)}{P(k = i)} = \frac{\lambda}{k + 1}, \ k \geq 0$$

so that we have a way to calculate $P(k = i + 1)$ given $P(k = i)$. If we select a uniform sample, $U[0, 1)$, we can sum the probability of selecting $k = 0, 1, 2, \ldots$ until we exceed U. This then is returned as the Poisson sample. We used the same approach above in Sect. 3.3 and in Fig. 3.12.

In C the algorithm is,

```
 1   int poisson_inv(double l) {
 2        double u=uniform(), s, p=exp(-l);
 3        int k=0;
 4        s = p;
```

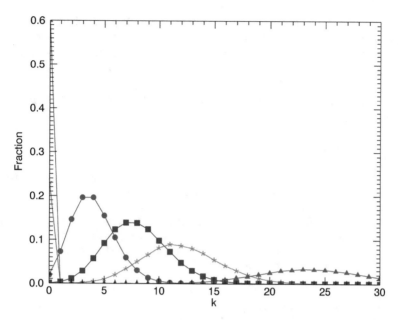

Fig. 3.13 Histograms of Poisson distributed samples from Poisson(4) (circle), Poisson(8) (square), Poisson(12) (star), and Poisson(24) (triangle). All samples were drawn using the multiplication method. Note that for $\lambda = 12$ and $\lambda = 24$ while the general shape of the histogram is correct there are a large number of $k = 0$ samples which should not be present. These are due to floating-point rounding errors implying that the multiplication algorithm should only be used for $\lambda < 10$

```
 5        while (u > s) {
 6            k++;
 7            p *= 1/k;
 8            s += p;
 9        }
10        return k;
11    }
```
(poisson.c)

where we increment a sum in s and the next product in p, both of which are initialized to the $k = 0$ value (lines 2 and 4). We only need one uniform (line 2) and then use the recurrence relation to accumulate probabilities for each k (lines 7 and 8) until we exceed the uniform sample. Returning k then gives us the Poisson(λ) sample.

This algorithm, like the multiplication algorithm, runs in time proportional to λ, but seems slightly less susceptible to round-off errors. Still, if λ is large, run time will become an issue, as we will see below. We can do better still.

Kemp (and Kemp) in [11] present an algorithm that is the end product of several other algorithms by Kemp and other authors. The specific details of the approximations used are beyond our present purposes, they are detailed in the paper, but in essence, the algorithm uses a search starting at the model max, as Devroye also stated is a way to speed up the algorithm. Once the modal point is determined and the proper search direction is selected, a quick loop selects the proper k to return. Like the inversion method, this method only uses a single uniform sample, which is what we were hoping to do. The code given in [11] is in Fortran 77. If we rework the code for C we get,

```
 1  int poisson_kemp(double l) {
 2      int i,mmin;
 3      double c,c1,c2,c3,d,d1,d2,d3,d4,e,e1,e2,f1,f2,f3,h;
 4      double qq,u,a,aa,g,m,p,q,rm,sp,sq;
 5
 6      d=1./2.; e=2./3.; c1=1./12.; c2=1./24; c3=293./8640.;
 7      d1=23./270.; d2=5./56.; d3=30557./649152.; e1=1./4.;
 8      e2=1./18.; f1=1./6.; f2=3./20.; f3=1./20.; h=7./6.;
 9      c=0.3989423; d4=34533608./317640015.;
10
11      u = uniform();
12      m = (int)(l+d);
13      rm = 1.0/(double)m;
14      a = l-m;
15      g = c/sqrt((double)m);
16      aa = a*a;
17      p = g*(1.0-c1/(m+c2+c3*rm));
18      q = p*(1.0-d*aa/(m+a*(e+a*(e1-e2*rm))));
19
20      if (u < d)
21          return poisson_kemp_downward(l,u,q,m);
22      if (u > (d+h*g))
23          return poisson_kemp_upward(l,u,q,m);
24
25      sp = d+g*(e-d1/(m+d2+d3/(m+d4)));
26      sq = sp-a*p*(1.0-f1*aa/(m+a*(d+a*(f2-f3*rm))));
27
28      if (u > sq)
29          return poisson_kemp_upward(l,u,q,m);
30      return poisson_kemp_downward(l,u,q,m);
31  }
```
(poisson.c)

which is not easily interpreted without recourse to the Kemp paper. Constants for approximations derived in the paper are computed in lines 6 through 9. We place them here to keep the code compact. Certainly they could be replaced with #define statements to avoid calculating them again on each call. Indeed, the original Fortran 77 code includes checks to see if the function is being called again with the same λ to avoid unnecessary re-initialization. We have removed those checks here to simplify

as much as possible and to avoid offending our modern sensibilities by using goto
statements in C. That said, there is a certain elegance to the compactness of the
original Fortran code, goto statements included.

The goal of the code in poisson_kemp is to determine whether to
search up or down leading to calls to either poisson_kemp_downward or
poisson_kemp_upward,

```
 1  int poisson_kemp_downward(double l, double u,
 2                            double q, int m) {
 3      int i, mmin;
 4      if (u < q)
 5          return m;
 6      mmin=m-1;
 7      for(i=0; i<mmin; i++) {
 8          u=u-q;
 9          q=(m-i)*q/l;
10          if (u < q)
11              return mmin-i;
12      }
13      return 0;
14  }
15  int poisson_kemp_upward(double l, double u,
16                          double q, int m) {
17      int i;
18      u=1.0-u;
19      for(i=m+1; i<=m+m+30; i++) {
20          q=q*l/(double)i;
21          if (u < q)
22              return i;
23          u = u-q;
24      }
25  }
```
(poisson.c)

How does the Kemp algorithm compare to the inversion and multiplication
algorithms? For small λ they all generate equivalent histograms when using the
same uniform source. This is also true for the inversion and Kemp algorithms when
λ is larger as seen in Fig. 3.14 where the histograms overlap completely.

We also see that the Kemp algorithm is robust to the source uniform samples
by plotting the histograms of Poisson(14) for one million samples each from the
same set of uniform generators used in Fig. 3.10 with the same seed values. The
overlapping plot is shown in Fig. 3.15.

Finally, comparing the mean runtime to generate one million samples over five
runs each of the inversion and Kemp algorithms for $\lambda \in \{4, \ldots, 400\}$ gives us
Fig. 3.16 clearly demonstrating the superior runtime performance of the Kemp
algorithm and the expected linear in λ runtime of the inversion algorithm.

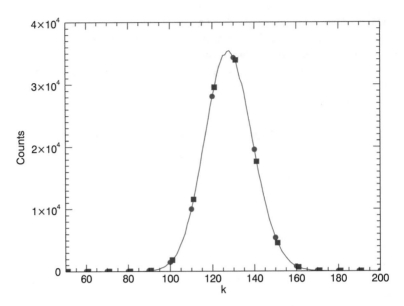

Fig. 3.14 Histograms for $\lambda = 128$ generated using the Kemp algorithm (circles) and inversion algorithm (squares) showing that both algorithms give identical results when using the same uniform sample source

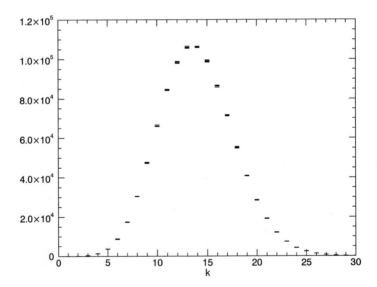

Fig. 3.15 Histograms for Poisson(14) using the Kemp algorithm and the same uniform source generators as in Fig. 3.10 with the same seed values

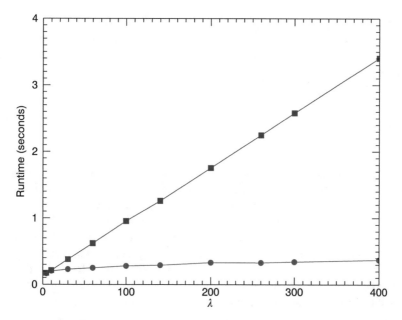

Fig. 3.16 Mean runtime ($n = 5$) to generate one million samples each as a function of λ for the inversion (square) and Kemp (circle) algorithms

3.7 Chapter Summary

In this chapter we developed approaches to draw random samples from a set of common distributions. For each distribution we implemented an algorithm in C, often more than one, and compared performance when using different uniform pseudorandom generators. We strove to select algorithms that are performant and that minimize the number of uniform samples necessary as inputs, ideally using a one-to-one mapping.

Exercises

3.1 Create a program to transform a file of random bytes, such as those of Chap. 1, into a file of C doubles for use with the test programs of this chapter. How do sources of truly random bytes compare to the output of the uniform pseudorandom generators?

3.2 Several times in this chapter we saw that even the terrible RANDU LCG generator still produced output samples that were acceptable. Discuss why this might be so in these cases.

3.3 Another commonly used distribution is the χ_k^2 distribution for k degrees of freedom. This distribution is related to the gamma distribution through $\chi_k^2 \sim$ Gamma$(\frac{k}{2}, \frac{1}{2})$. Define functions to sample from χ_k^2 using both the Ahrens and Marsaglia algorithms and compare histograms for different k values. Compare with the PDF defined by,

$$
y = \frac{1}{2^{\frac{k}{2}} \Gamma\left(\frac{k}{2}\right)} x^{\frac{k}{2}-1} e^{-\frac{x}{2}}
$$

3.4 ^{137}Cs is radioactive and decays with a half-life of 30.17 years so that the probability of any one nucleus decaying in one second is about 3.12×10^{-7}. A 1 μg sample has about 4.4×10^{15} nuclei so that approximately 3.2×10^6 nuclei will decay in 1 s. Write a simulation program that will output samples of the number of decays measured per given time interval in seconds. Run the simulation for several thousand samples for each interval and plot the histogram. Use intervals of seconds, days, and weeks. Ensure that it follows a Poisson distribution. Note, for a Poisson process, the parameter is the mean number of events expected per unit interval, t. Scale the per second rate appropriately for t in seconds before sampling. **

3.5 Simulation is a primary use for sampling from nonuniform distributions. Create a program to generate a population of people where each person is described by samples from a particular distribution. Specifically, define a person by attributes for gender, age, height, weight, education level, income, and likelihood of having lung cancer, breast cancer, and diabetes. Note, some of these distributions are contingent on other properties. For example, gender is a strong determinant of risk for breast cancer. Also, educational level affects income. For the United States, a good source of information about what these distributions look like is the Census Bureau (https://www.census.gov/). ***

References

1. Box, George EP, and Mervin E. Muller. "A note on the generation of random normal deviates" The annals of mathematical statistics 29.2 (1958): 610–611.
2. Marsaglia, George, and Thomas A. Bray. "A convenient method for generating normal variables." Siam Review 6.3 (1964): 260–264.
3. Devroye, Luc. Non-Uniform Random Variate Generation (1986): 524ff.
4. Marsaglia, George, and Wai Wan Tsang. "A simple method for generating gamma variables." ACM Transactions on Mathematical Software (TOMS) 26.3 (2000): 363–372.
5. Ahrens, Joachim H., and Ulrich Dieter. "Computer methods for sampling from gamma, beta, poisson and binomial distributions." Computing 12, no. 3 (1974): 223–246.
6. Ahrens, Joachim H., and Ulrich Dieter. "Generating gamma variates by a modified rejection technique." Communications of the ACM 25, no. 1 (1982): 47–54.
7. Marsaglia, George. "The squeeze method for generating gamma variates." Computers and Mathematics with Applications 3, no. 4 (1977): 321–325.

8. Kneusel, Ronald T. and Mozer, Michael C. "Improving Human-Machine Cooperative Visual Search With Soft Highlighting." ACM Trans. Appl. Percept., 15 (2017): 3:1–3:21.

9. Bevington, Philip T. "Data reduction and error analysis for the physical sciences." (1969).

10. Knuth, D.E.. Seminumerical Algorithms. The Art of Computer Programming (2nd ed.), Vol. 2, Addison-Wesley, Reading, MA (1981).

11. Kemp, C. D., and Adrienne W. Kemp. "Poisson random variate generation." Applied Statistics (1991): 143–158.

Chapter 4
Testing Pseudorandom Generators

Abstract Testing pseudorandom number generators is not quite as straightforward as it might seem. In this chapter we consider classical tests of randomness and apply them to the generators discussed in Chap. 2. Next we investigate two popular test suites: dieharder (based on the older DIEHARD) and TestU01, and one quick test program (ent). These test suites are the benchmarks against which researchers generally measure new algorithms.

When researchers develop new algorithms for generating pseudorandom sequences of numbers they need some sort of common framework in which to evaluate those algorithms. This is where the techniques of this chapter come into play. We will look at the classical tests initially applied to pseudorandom generators and implement some of them so we can test the generators of Chap. 2. However, while genuinely useful, it is not enough, from a community perspective, to have a loose collection of tests. Over time there emerged test suites which are the de facto standards against which new algorithms are measured. We will detail and use two of them in this chapter: the dieharder tests of Brown [1], which are based on the older DIEHARD tests of Marsaglia, and the TestU01 suite of L'Ecuyer and Simard [2]. We will conclude with a look at the *ent* program that is simpler and useful for testing files for randomness.

4.1 Classical Randomness Tests

There are many ways of assessing the quality of a pseudorandom generator or a file of numbers created by a pseudorandom or random process. The traditional set of tests is presented at length in [3]. We will discuss and implement some these, or variants of them, and apply them to the generators of Chap. 2 as well as the (hopefully) truly random sequences of Chap. 1.

We will implement two hypothesis tests: the χ^2 test, also called the frequency test, and the Kolmogorov-Smirnov or KS-test. To these we will add several

© Springer International Publishing AG, part of Springer Nature 2018   115
R. T. Kneusel, *Random Numbers and Computers*,
https://doi.org/10.1007/978-3-319-77697-2_4

empirical tests: serial test, gap test, maximum-of-t test, permutation test, the serial correlation test. All of these are from Knuth [3]. Additionally, we will implement one bit-level test, the random excursions test, which is from the NIST test suite [4]. Space considerations force us to otherwise ignore the NIST test suite. Readers are encourage to examine the suite on their own.

Each of the classical randomness tests follows the same general format. Many will make use of the χ^2 statistic so we will implement it first and then refer back to the code in the implementations of the other tests. In all cases, the test will be implemented as a C program that reads from the command line. The input source will be a file, sometimes the file will be a binary file of double precision values in the range [0, 1) and other times the file will be interpreted as a stream of bits, bytes or integers of some size.

We implement each test first, with source code, and then in Sect. 4.2 apply them to see how our generators and random sequences stand up.

4.1.1 χ^2 Test

The χ^2 test statistic is intuitively derived, following Knuth, from a set of n measurements, each falling into one of k possible categories, with expected probabilities p_k. If n is large (something we will come back to later), then the expected number of results falling into category k for n repeats of the measurement will be np_k, i.e., n times the probability, p_k, of event k. If we tally the number of events actually falling into category k, called x_k, we can sum the deviation across categories, squaring to avoid caring about whether we are above or below the expected value, to give,

$$v = \sum_{1 \le i \le k}^{k} \frac{(x_i - np_i)^2}{np_i}$$

where we need to scale each deviation $(x_i - np_i)^2$ by the expected number, np_i, to weigh the deviations by how important they are. If category i is expected 10,000 times in n trials a deviation of 100 matters less than a deviation of 100 in a category that is expected to occur only 200 times in n trials.

The value, v, when combined with a number of degrees of freedom (read $k - 1$), gives us a p-value, a probability against which we can decide to accept or reject the randomness of the measurements. We will see in the code below how to calculate this p value. One item is worth remembering here. We are looking for p-values that are not too small, the typical case, nor too large. Either one indicates a significant departure from randomness. Knuth gives ranges, we will simplify them and use <0.05 and >0.95 as our indicators. If a p-value is between these two limits, $0.05 < p < 0.95$, we will claim this as evidence for the randomness of the sequence.

The full test program will be given here. The χ^2 portion, especially the calculation of the p-value, will be used in several of the empirical tests as well. Our implementation is really the frequency test. We will generate a histogram of the occurrences of values in a number of bins from [0, 1) under the reasonable assumption that with enough values a truly random sequence will approach a constant number in each bin. The χ^2 test will decide if this assumption is valid for the given sequence.

We start with the driver program. This skeleton will be duplicated for all our tests with appropriate changes when necessary. The code is,

```
1  int main(int argc, char *argv[]) {
2      FILE *g;
3      int typ;
4      double v, p;
5
6      typ = atoi(argv[1]);
7      g = fopen(argv[2],"r");
8      v = chisq(typ, g);
9      p = chisq_pval(v, 255);
10     printf("\nchisq  = %0.8f, p-value = %0.8f\n\n", v, p);
11     fclose(g);
12 }
```
(chisq.c)

where, to save space, we have removed the lines that display how to use the program when the program is run without arguments. The first argument is an integer in $\{0, 1, 2, 3\}$ indicating the C data type of the file of values that we are testing: byte, unsigned 32-bit integer, unsigned 64-bit integer, or double precision floating-point value $[0, 1)$. The second argument is the name of the file. The program itself produces no output other than the χ^2 and its associated p-value.

Notice that lines 8 and 9 call two additional functions. The chisq function will perform the frequency test and return the χ^2-statistic, v. The chisq_pval function will turn the test statistic into a probability. This function will be used any time we need to change a χ^2 value into a probability. It is a software form of the old χ^2 table we learned about in kindergarten or at some point after that. Let's look first at chisq,

```
1  double chisq(int typ, FILE *g) {
2      int n=0,i,b,bins[256];
3      double v=0.0, p=1.0/256.0;
4
5      memset((void*)bins, 0, 256*sizeof(int));
6      while (1) {
7          b = next(typ,g);
8          if (b == -1) break;
9          bins[b]++;
10         n++;
11     }
12     for(i=0; i<256; i++)
13         v += ((bins[i]-n*p)*(bins[i]-n*p))/(n*p);
14     return v;
15 }
```
(chisq.c)

This function takes the input file (g) and generates a histogram with a fixed size of 256 bins. Line 2 defines the histogram vector, bins, and line 3 assigns the probability of landing in any one of the bins to p. This probability is, of course, the same for all bins so it is simply 1/256. Line 5 uses a library function to set all the bins to an initial value of zero.

Lines 6 through 11 read all the values from the file and fill in the histogram. To understand what this unexpected, seemingly infinite, loop is doing we need to define the next function. All of our test programs will use this next function or some variant of it. For the moment, all we need know is that next returns the next value from the input file we are testing. In this case it is returning the actual bin into which the next value would fall. This lets us build the histogram trivially (line 9). Line 8 is the exit from the infinite loop. When there are no more values in the input file, next returns −1. This works for us because all our values are either floating-point [0, 1) or unsigned integers. Line 10 counts the number of values read (n).

Lines 12 and 13 are the actual calculation of the χ^2 test statistic. We take the difference between the number of values in bin i and the expected number. If the probability of being in bin i is p (line 3) when the file is random, then the expected number of values in bin i is np and this is true of all bins. The denominator is our scale factor and we loop over all the bins to generate v which is returned.

Let us consider the function next,

```
 1 int next(int typ, FILE *g) {
 2     uint8_t b8;
 3     uint32_t b32;
 4     uint64_t b64;
 5     double d;
 6     switch (typ) {
 7         case 0:
 8             if (fread((void*)&b8, sizeof(uint8_t), 1, g)
                    == 0)
 9                 return -1;
10             return (int)b8;
11             break;
12         case 1:
13             if (fread((void*)&b32, sizeof(uint32_t), 1, g)
                    == 0)
14                 return -1;
15             return (int)(b32>>24);
16             break;
17         case 2:
18             if (fread((void*)&b64, sizeof(uint64_t), 1, g)
                    == 0)
19                 return -1;
20             return (int)(b64>>56);
21             break;
22         case 3:
```

```
23              if (fread((void*)&d, sizeof(double), 1, g) ==
                0)
24                  return -1;
25              return (int)(256*d);
26              break;
27          default:
28              break;
29      }
30      return -1;
31 }
```
(chisq.c)

which is somewhat long but really just the same bit of code repeated. The first
argument is typ which decides how we should read and interpret the next value
from the input file, g. If typ==0 we are working with bytes. We read the next byte
from the file in line 8 and use the return value of fread to return -1 when there are
no more bytes to read. Line 10 simply returns the byte because we cleverly chose to
use 256 bins in our histogram so the value read is the index.

We repeat the above for 32-bit unsigned integers (line 12) where we shift the
integer down by 24 bits to leave only the top 8 bits which, as a byte, are our index.
Line 17 starts the case for 64-bit unsigned integers where we now shift down 56 bits
to again leave the top 8 bits. Finally, case 3 (line 22) reads a double precision
floating-point number. Multiplying this number by 256 and taking the floor (line
25) returns the bin number indicating where the floating-point value read resides.

We now need to give the code for the calculation of the p-value, the function
chisq_pval,

```
1 double chisq_pval(double cv, int dof) {
2     double k = (double)dof * 0.5;
3     double x = cv * 0.5;
4
5     if ((cv < 0) || (dof < 1))
6         return 0.0;
7     if (dof == 2)
8         return exp(-x);
9     return igf(k,x)/tgamma(k);
10 }
```
(chisq.c)

The inputs are the χ^2 statistic (cv) and the associated degrees of freedom (dof)
represented here by 255 since we have 256 bins. Lines 2 and 3 divide these values
by 2. We see in Problem 3.3 that the χ^2 distribution is derivable from the gamma
distribution which partially explains dividing each input by two.

Lines 5 through 8 are simple checks. If the degrees of freedom is only two, the
decaying exponential returns the value that we want. Otherwise, line 9 is used. This
line uses the C standard library function, tgamma, and an implementation of the
incomplete gamma function,

```
 1  double igf(double s, double z) {
 2      double sc=(1.0/s), sum=1.0;
 3      double num=1.0, denom=1.0;
 4      int i;
 5
 6      if (z < 0.0)
 7          return 0.0;
 8      sc *= pow(z,s);
 9      sc *= exp(-z);
10      for(i=0; i<60; i++) {
11          num *= z;
12          s++;
13          denom *= s;
14          sum += (num/denom);
15      }
16      return sum*sc;
17  }
```
(chisq.c)

to return the *p*-value. The incomplete gamma function is an integral,

$$\Gamma(s, z) = \int_z^\infty t^{s-1} e^{-t} dt$$

which the function `igf` approximates with the sum in lines 10 through 15. The number of iterations was set empirically as a trade-off between accuracy and round-off errors.

The output of this program is the χ_k^2 value and its associated *p*-value. We use the *p*-value to make our judgment. If it is too small (<0.05) or too large (>0.95) we have evidence that the samples should probably not be considered random. Knuth uses ranges and treats <0.05 or >0.95 as suspect. Certainly, any *p*-value in the range [0.25, 0.75] is a strong indicator of randomness in the sequence. The quality of this judgement is highly dependent upon the number of values examined. A small file will lead to unreliable results. What is a small file? That is a good question. In practice, one should try to use as large a file as possible.

4.1.2 Kolmogorov-Smirnov Test

The Kolmogorov-Smirnov test is a nonparametric test that measures the difference between two probability distributions. For our purposes, the first distribution is the uniform distribution we would expect from a truly random generator and the second is the empirical distribution generated from the actual sample data.

Specifically, we need to build the cumulative distribution functions to return the probability that a sample value, *x*, is less than *X* for *X* in [0, 1]. For a uniform distribution the cumulative distribution function is simply the line $y = x$ since the probability that a uniform sample $x \sim [0, 1)$ is less than $X = 0.7$ is 0.7. Therefore, if we can generate the empirical cumulative distribution function we can generate deviations from the uniform cumulative distribution function and this is precisely what we need to generate the KS-test statistic.

A simple driver program,

```
1  int main(int argc, char *argv[]) {
2       FILE *g;
3       int typ;
4       double v, p;
5       typ = atoi(argv[1]);
6       g = fopen(argv[2],"r");
7       v = kstest(typ, g);
8       fclose(g);
9  }
```
(kstest.c)

is sufficient for this test. It gets the data type (`typ`) and the file name from the command line and then opens the file (`g`). The magic happens in `kstest`,

```
1  double kstest(int typ, FILE *g) {
2       int i,k,n;
3       size_t w = sizeof(uint8_t);
4       double d,dp=-1.0,dm=-1.0,kp,km, z[EMAX], *p;
5
6       memset((void*)z, 0, EMAX*sizeof(double));
7       if (typ==1) w = sizeof(uint32_t);
8       if (typ==2) w = sizeof(uint64_t);
9       if (typ==3) w = sizeof(double);
10      fseek(g, 0, SEEK_END);
11      n = ftell(g) / w;
12      fseek(g, 0, SEEK_SET);
13      if (n > MAX_VALUES) n = MAX_VALUES;
14
15      p = (double *)malloc(n*sizeof(double));
16      for(i=0; i<n; i++)
17          p[i] = next(typ,g);
18
19      for(i=0; i<EMAX; i++)
20          for(k=0; k<n; k++)
21              if (p[k] <= ((double)i/EMAX)) z[i]++;
22      for(i=0; i<EMAX; i++) {
23          d = z[i]/n - (double)i/EMAX;
24          if (d > 0)
25              if (d > dp)
```

```
26                    dp = d;
27            if (d < 0)
28                if (fabs(d) > dm)
29                    dm = fabs(d);
30        }
31
32        kp = dp*sqrt(n);
33        km = dm*sqrt(n);
34        free(p);
35        printf("kp=%0.6f, km=%0.6f\n", kp,km);
      (kstest.c)
36  }
```

where lines 6 through 13 are configuration to read the sample file once. The empirical cumulative distribution function is stored in z with EMAX bins partitioning the range [0, 1]. Here EMAX is 100 and z is set to zero everywhere via line 6. Lines 7 through 13 determine how many values are actually stored in the input file (n). If this number exceeds MAX_VALUES we only use MAX_VALUES. Here MAX_VALUES is 1000. Since we need to maintain the values in memory we do not want to make MAX_VALUES too large nor too small, lest the test results not be reliable. Lines 15 through 17 read the samples into memory (p) by calling next.

The actual KS-test is in lines 19 through 33. Lines 19 through 21 build the empirical cumulative distribution function in z. For the current bin number (i) we make a pass through the samples looking for every sample that is less than or equal to the value of that bin for the uniform cumulative distribution function (line 21). When one is found we increment the count in z so that when complete z is a histogram approximating the cumulative distribution function.

Lines 22 through 30 determine the signed difference, per bin, between the empirical distribution (scaled by n, the number of samples) and the expected value of a uniform distribution (line 23). We then track the largest positive (lines 24–26) and negative differences (lines 27–29). At the end of this loop we know the largest positive and negative deviation. Lines 32 and 33 transform the deviations into the KS test statistics, K_n^+ (kp) and K_n^- (km). The $\sqrt{n}$ factor is present to keep the test independent of the standard deviation (see Knuth [3]).

The next function is similar in form to the one used in Sect. 4.1.1 but in this case we want a double precision floating-point value regardless of the input source type. For bytes, divide by 256.0. For unsigned 32-bit integers we divide by 2^{32}. For unsigned 64-bit integers we cheat somewhat and shift the 64-bit value down by 32 bits before dividing by 2^{32}. Finally, for floating-point input we simply return the value read.

So, how does one interpret the output of this test? Knuth gives in Table 2 [3, page 48] values for K_n^+ and K_n^- for different thresholds. Using this table and $n = 1000$ we calculate that if $0.1548 < K_n^+, K_n^- < 1.2186$ holds then we can say we have evidence for randomness.

4.1.3   Serial Test

The serial test looks for correlations between x_n and x_{n+1}. It does this by generating a two-dimensional histogram of how often each possible pair of x values occurs in the input file and then applies the χ^2 test to look for deviations from uniformity. Here we use a relatively small number of bins, 16 in fact, so that there are enough counts to make the χ^2 test meaningful. The input file should have at minimum 10,000 values. For this test, the code is quite similar to that of Sect. 4.1.1. The function next needs to return a bin number for the input float [0, 1). For example, if the input type is byte, then (int)(BINS*(b8/256.0)) will return a bin number, [0, BINS). A similar modification is necessary for the other input data types.

The test code itself uses a simple driver,

```
 1  #define BINS 16
 1  int main(int argc, char *argv[]) {
 2      FILE *g;
 3      int typ;
 4      double v, p;
 5      typ = atoi(argv[1]);
 6      g = fopen(argv[2],"r");
 7      v = chisq(typ, g);
 8      p = chisq_pval(v, BINS*BINS-1);
 9      printf("\nchisq  = %0.8f, p-value = %0.8f\n\n", v, p);
10      fclose(g);
11  }
```
(serial_test.c)

to get the data type (typ) and open the input file (g). It then calls chisq and the chisq_pval function of Sect. 4.1.1 where the degrees of freedom is BINS squared minus one because we are using a two-dimensional histogram. All the action is in chisq,

```
 1  double chisq(int typ, FILE *g) {
 2      int n=0,i,j,b0,b1,bins[BINS*BINS];
 3      double v=0.0, p=1.0/(BINS*BINS);
 4  
 5      memset((void*)bins, 0, (BINS*BINS)*sizeof(int));
 6      while (1) {
 7          b0 = next(typ,g);
 8          if (b0 == -1) break;
 9          b1 = next(typ,g);
10          if (b1 == -1) break;
11          bins[BINS*b0+b1]++;
12          n++;
```

```
13  |      }
14  |      for(i=0; i<(BINS*BINS); i++)
15  |          v += ((bins[i]-n*p)*(bins[i]-n*p))/(n*p);
16  |      return v;
17  | }
```
(serial_test.c)

where we zero the histogram (line 5) and read pairs of values (lines 6–13) each time incrementing the appropriate bin of the histogram. Note, instead of a two-dimensional C array we have used a single vector and applied the equation mapping two indices (b0, b1) to a location in memory (BINS*b0+b1). We count how many *pairs* we have read in n.

Lines 14 and 15 calculate the χ^2 statistic. The count in each bin is compared to the expected count which is simply np for $p = 1/256$ (line 3) since we expect each pair to appear equally often. The statistic is returned and the main program finds the p-value and reports it. As before, we use p-values of <0.05 or >0.95 as evidence against randomness.

This program looks at successive pairs, (x_n, x_{n+1}), though it could be expanded to look at triplets or even higher dimensions. Thought of this way, it is similar to the concept of equidistribution or volume filling. Problem 4.2 asks the reader to extend the code to three dimensions and then apply it to the output of the RANDU generator (see Sect. 2.2). Note, it is important that disjoint sets are used. Using (x_0, x_1) and (x_2, x_3) is correct but using (x_0, x_1) and (x_1, x_2) would not be correct because x_1 would be counted twice and violate the independence condition.

4.1.4 Gap Test

The first three tests in this section generate histograms of the frequencies with which values or pairs of values occur in the sequence being tested. The gap test instead looks at the distribution of runs within a particular range. For example, if the range is $[0.0, 0.5)$ then we want to track the number of times a sequence of samples falls outside this range before the next value that falls within this range. The input to this test should be 100 million or more samples to produce reliable results.

For example, if the sequence of values (showing only one decimal) is,

$$0.4, 0.6, 0.8, 0.7, 0.2, 0.8, 0.6, 0.4, 0.8, 0.9, 0.7, 0.1$$

then there is a run of three (0.6, 0.8, 0.7), a run of two (0.8, 0.6), and another run of three (0.8, 0.9, 0.7). If we track the lengths of the runs found when making a pass through the samples we have a histogram against which we can apply the χ^2 test. The probabilities for each bin are,

$$p_0 = \beta - \alpha, \quad p_i = p(1-p)^i, \quad p_t = (1-p)^t$$

for a search range $[\alpha, \beta)$. Here t is the number of bins, i.e., the maximum run length tracked. Any run above t is simply put in the t bin. For the example above, $\alpha = 0$ and $\beta = 0.5$.

For this test we need to interpret the input values as double precision floats so the next function will be of the same form as in Sect. 4.1.3. A simple driver program runs the test for two common ranges, runs below the mean ($[0, 0.5)$), and runs above the mean ($[0.5, 1.0)$),

```
1  int main(int argc, char *argv[]) {
2      FILE *g;
3      int typ;
4      double v, p;
5
6      typ = atoi(argv[1]);
7      g = fopen(argv[2],"r");
8      p = gap_test(0.0, 0.5, typ, g, &v);
9      printf("\nruns below the mean, chisq  = %0.8f, p-value
           = %0.8f\n", v, p);
10     fclose(g);
11     g = fopen(argv[2],"r");
12     p = gap_test(0.5, 1.0, typ, g, &v);
13     printf("runs above the mean, chisq  = %0.8f, p-value
           = %0.8f\n\n", v, p);
14     fclose(g);
15 }
   (gap_test.c)
```

where we use a data type argument (typ) and a file of samples (g). Line 8 performs the test for runs below the mean. The χ^2 value is returned in v and the associated p-value is in p. Similarly, line 12 performs the test for runs above the mean.

The gap_test function does the work,

```
1  double gap_test(double a, double b, int typ, FILE *g,
   double *v) {
2      int n=100000000,s=0,i,r,count[T];
3      double p=b-a, u, prob[T];
4
5      memset((void*)count, 0, T*sizeof(int));
6
7      while (s < n) {
8          r = 0;
9          u = next(typ,g);
10         if (u == -1) break;
11         while ((u < a) || (u > b)) {
12             r++;
13             u = next(typ,g);
14             if (u == -1) break;
15         }
```

```
16            if (u == -1) break;
17            if (r >= T) count[T]+=1; else count[r]+=1;
18            s++;
19        }
20
21        for(i=0; i<T; i++)
22            prob[i] = p*pow(1.0-p, i);
23        prob[0] = p;
24        prob[T-1] = pow(1.0-p, T);
25
26        *v = 0;
27        for(i=0; i<T; i++)
28            *v += (count[i]-s*prob[i])*(count[i]-s*prob[i])/
                  (s*prob[i]);
29        return chisq_pval(*v,T-1);
30 }
```
(gap_test.c)

where T is the number of bins, here $T = 32$.

The arguments are the range, $[a, b)$, the data typ, the source file, g, and the output χ^2 value, v. We process at most $n = 100,000,000$ samples storing the gap lengths in count which we initialize to all zeros (line 5).

There are three sections. Lines 7–19 scan through the samples building the histogram of gap lengths. Lines 21–24 calculate the bin probabilities so we can apply the χ^2 test. Finally, lines 26–29 calculate the χ^2 statistic and return it in v with the p-value (line 29).

Let's look closely at the first section (lines 7–19) as this is the core of the test. The index into the bins (stored in count) is r initialized to zero. The loop reads a floating-point value from the samples file (line 9) and starts a loop (line 11). Recall, if the return value of next is -1 all the samples have been read. The test condition of the while loop checks to see if the sample is outside of the desired range. If so, the bin index is incremented (line 12) and the next sample is read (line 13). Once the condition is false we have a sample in the desired range. This marks a gap so we need to increment the proper bin to record an occurrence of that gap size (line 17). Line 18 counts how many gaps we have found. This is used to control the outer loop (line 7) which locates up to n gaps.

The second section (lines 21–24) generates a vector of probabilities corresponding to the likelihood that a gap will fall into a particular bin. We first fill in all the elements of the vector using the general equation for the probability and then reset the first and last bin with the proper values (lines 23 and 24).

The χ^2 statistic is calculated in lines 26–28. We accumulate the squared differences between the actual number of gaps found in a bin (count) and the expected number which is the number of gaps found (s) times the probability for that bin (prob[i]) divided by the expected number. The statistic is returned in v and the p-value from line 29. There are T bins so we have $T - 1$ degrees of freedom.

Since this test returns a χ^2 p-value, we use the same decision thresholds as before. If $p < 0.05$ or $p > 0.95$ we reject the claim that the sequence is random.

4.1.5   Maximum-of-t Test

The maximum-of-t test applies the frequency test to the largest value of sets of samples. Generating the data for the test is straightforward: find the largest value in t consecutive samples and store this value raised to the t power in a histogram of a set number of bins, N. Then repeat for the next set of t samples until all samples have been acquired. Each maximum value is a measurement and when the entire input file has been processed we will have n measurements. Here $n > 1,000,000$ is a reasonable value to give a meaningful result.

We have every expectation that the histogram generated in this way will be uniform if the input sequence is random so we know that the probability associated with any bin is simply $p = 1/N$ and the expected number in each bin will be np. From this we can generate a χ^2 statistic and its associated p-value. As before with the χ^2 test, we can consider p-values <0.05 or >0.90 as evidence that the input sequence is not random.

If m is the maximum value of a set of t samples, why are we tracking m^t and not just m? The probability that any sample is $<x$ for some x in $[0, 1)$ is simply x if the samples are uniformly distributed (i.e. random). This is also true of the next sample, the samples are (assumed) independent. Therefore, the probability that in t samples the maximum value is $<x$ is the product of the probabilities that each of the samples is $<x$ which is simply x giving $xxxx\ldots = x^t$. It is this value that we would expect to be uniformly distributed over $[0, 1)$ and to which we apply the χ^2 test.

The `next` function this test uses must return a double precision floating-point value regardless of the data type of the input file. It will also use the `chisq_pval` function of Sect. 4.1.1. The driver is,

```
 1  int main(int argc, char *argv[]) {
 2      FILE *g;
 3      int typ, t;
 4      double v, p;
 5
 6      t = atoi(argv[1]);
 7      typ = atoi(argv[2]);
 8      g = fopen(argv[3],"r");
 9      v = max_of_t(t, typ, g);
10      p = chisq_pval(v, NBINS-1);
11      printf("\nmax-of-%d chisq  = %0.8f, p-value = %0.8f\n
           \n", t, v, p);
12      fclose(g);
13  }
```
(max_test.c)

The test takes three inputs: the t value, the data type (typ), and the name of the samples file. Lines 6–8 read the command line arguments. Line 9 calls max_of_t to perform the test and return the χ^2 value. Lines 10 and 11 find the associated p-value and report it. For this test we fix the number of histogram bins (NBINS) at 20.

The test itself is,

```
 1  double max_of_t(int t, int typ, FILE *g) {
 2      int i,n=0,bins[NBINS];
 3      double b,m, v=0.0, p=1.0/(double)NBINS;
 4
 5      memset((void*)bins, 0, NBINS*sizeof(int));
 6      while (1) {
 7          m = -1.0;
 8          for(i=0; i<t; i++) {
 9              b = next(typ,g);
10              if (b == -1) break;
11              if (b > m) m=b;
12          }
13          if (b == -1) break;
14          n++;
15          bins[(int)(NBINS*pow(m,t))]++;
16      }
17      for(i=0; i<NBINS; i++)
18          v += ((bins[i]-n*p)*(bins[i]-n*p))/(n*p);
19      return v;
20  }
```
(max_test.c)

where line 5 zeroes the histogram. We then start a loop (lines 6–16) to read t values from the sample file tracking the largest value read, m (lines 7–12). Line 14 increments the counter, how many sets have been read. Line 15 counts the occurrence of m^t by incrementing the proper bin of the histogram, bins. Simple multiplication by the number of bins works here because m is in $[0, 1)$ so m^t is also in $[0, 1)$. The proper bin is some fraction of the number of bins available, as an integer, which is what we get from (int)(NBINS*pow(m,t)). When all sets have been processed the loop exits and lines 17 and 18 calculate the χ^2 statistic as in the previous tests.

4.1.6 Serial Correlation Test

The serial correlation test calculates the correlation of the values in the input sample file. Specifically, it calculates the coefficient of the correlation between a value and the next value in the sequence. For a random sequence the correlation will be zero as the samples are independent. The actual statistic is,

$$C = \frac{n(U_0U_1 + U_1U_2 + U_2U_3 + \cdots + U_{n-2}U_{n-1} + U_{n-1}U_0) - (U_0 + U_1 + U_2 + \cdots + U_{n-1})^2}{n(U_0^2 + U_1^2 + U_2^2 + \cdots + U_{n-1}^2) - (U_0 + U_1 + U_2 + \cdots + U_{n-1})^2}$$

For uniform samples Knuth [3] states that 95% of the time C will (most likely) be between $[m - 2\sigma, m + 2\sigma]$ for $m = -1/(n - 1)$ and,

$$\sigma = \frac{1}{n - 1}\sqrt{\frac{n(n - 3)}{n + 1}}, \; n > 2$$

where n is the number of samples in the input sequence. In general, n should be several hundred or more.

In code we will want the samples in the range $[0, 1)$ so we need a version of next that returns double precision floating-point values regardless of the input data type.

The driver program is,

```
 1 int main(int argc, char *argv[]) {
 2     FILE *g;
 3     int typ,n;
 4     double c,m,s;
 5
 6     typ = atoi(argv[1]);
 7     g = fopen(argv[2],"r");
 8     c = corr(typ, g, &n);
 9     m = -1.0/(n-1.0);
10     s = (1.0/(n-1.0))*sqrt(n*(n-3.0)/(n+1.0));
11     printf("\ncorr = %0.5f (n=%d), expected 95%% CI=[%0.5
       f, %0.5f]", c, n, m-2*s, m+2*s);
12     if ((c >= m-2*s) && (c <= m+2*s))
13         printf(", test PASSED\n\n");
14     else
15         printf(", test FAILED\n\n");
16     fclose(g);
17 }
```
(corr_test.c)

where the only command line arguments are the source file data type (typ) and the file name assigned to g. Here we allow byte (0), unsigned 32-bit integer (1), unsigned 64-bit integer (2), and double precision float (3) as input data types.

Line 8 calculates the correlation statistic. Note that it also returns n, the number of samples. Lines 9 and 10 calculate the 95% confidence interval and line 11 reports the result. Lines 12–15 simply check to see if C is within the expected confidence interval for n samples and declares whether the test has passed or not.

The correlation function `corr` is straightforward,

```
 1 double corr(int typ, FILE *g, int *n) {
 2        double c,ui,u0,u1;
 3        double sum=0,ssum=0,sprod=0;
 4
 5        *n = 1;
 6        ui = next(typ,g);
 7        if (ui==-1) return -1;
 8        u0 = ui;
 9        sum = u0;
10        ssum = u0*u0;
11        while (1) {
12              u1 = next(typ,g);
13              if (u1==-1) break;
14              sum += u1;
15              ssum += u1*u1;
16              sprod += u0*u1;
17              u0 = u1;
18              (*n)++;
19        }
20        sprod += u1*ui;
21        return ((*n)*sprod-sum*sum)/((*n)*ssum-sum*sum);
22 }
   (corr_test.c)
```

with the `while` loop (lines 11–19) accumulating the sum of the samples (`sum`), the sum of the squares of the samples (`ssum`), and the sum of the products of the pairs (`sprod`). Line 20 adds in the final pair $(U_{n-1}U_0)$. The returned value then is the actual correlation statistic (line 21).

Unlike other tests in this section, this one reports success or failure based on the expected confidence interval. Regardless, intuition states that the correlation should be as close to zero as possible, either slightly positive or slightly negative.

4.1.7 Permutation Test

The permutation test examines the frequencies with which specific orderings of t consecutive samples occur. There are $t!$ orderings of t values and these orderings should occur equally often with a probability $p = 1/t!$ if the input samples are random. This test requires as input t which should be no more than 5. Additionally, if $t = 4$ or $t = 5$, then the number of input samples should be 100 million or more. The calculated statistic is χ^2.

The above begs the question: how to do we uniquely characterize the orderings of samples? One way to do this would be to build a table mapping a sequence of t elements to a particular numeric value, $[0, t)$. This, naturally, is not elegant and would quickly become cumbersome as t increases. For $t = 5$ we would already need a table of 120 elements and many machine cycles to determine which input sequence matches one of the table entries. Fortunately for us, there is a simpler way to uniquely characterize each of the possible $t!$ orderings. If we sort the sequence numerically we can generate a value, call it f, which is uniquely determined by the initial ordering of the samples. I.e., the act of sorting the samples will require shuffling entries and the sequence of shuffles will determine f.

The algorithm is as follows,

1. Set $f = 0, r = t$.
2. Find the largest value in the sequence from s_0 to s_r. Call this index i.
3. Swap the largest value and the s_r element.
4. Update $f, f = rf + i$.
5. Decrease r by one.
6. Repeat from Step 2 while $r > 1$.
7. Return f.

This algorithm will sort the elements in ascending order, a by-product we do not actually care about here, and calculate a *unique* integer value for $f, 0 \leq f < t$, that represents the initial ordering of the elements. It is this value that we use as an index into our histogram of measurements.

The samples for this test need to be double precision values so `next` must return these for any input data type. The driver program follows the expected form for this section,

```
 1  int main(int argc, char *argv[]) {
 2      FILE *g;
 3      int typ,t;
 4      double v, p;
 5
 6      t = atoi(argv[1]);
 7      if (t < 2) t=2;
 8      if (t > 5) t=5;
 9      typ = atoi(argv[2]);
10      g = fopen(argv[3],"r");
11      v = perm_test(t, typ, g);
12      p = chisq_pval(v, fact(t));
13      printf("\npermutation size=%d, chisq  = %0.8f, p-value
            = %0.8f\n\n", t, v, p);
14  }
    (perm_test.c)
```

where we ensure that t is in $[2, 5]$. The `perm_test` function returns the χ^2 statistic and we use `chisq_pval` to find the associated p-value. Line 13 reports the results. Any p in the range $[0.05, 0.95]$ is evidence for randomness.

Let's look at perm_test,

```
 1 double perm_test(int t, int typ, FILE *g) {
 2     int n=0,i,*bins,bn,f;
 3     double *b,x, v=0.0, p;
 4
 5     bn = fact(t);
 6     p = 1.0/(double)bn;
 7     b = (double*)malloc(t*sizeof(double));
 8     bins = (int*)malloc(bn*sizeof(int));
 9     memset((void*)b, 0, t*sizeof(double));
10     memset((void*)bins, 0, bn*sizeof(int));
11     while (1) {
12         for(i=0; i<t; i++) {
13             x = next(typ,g);
14             if (x==-1) break;
15             b[i] = x;
16         }
17         if (x==-1) break;
18         f = perm(t,b);
19         bins[f]++;
20         n++;
21     }
22     for(i=0; i<bn; i++)
23         v += ((bins[i]-n*p)*(bins[i]-n*p))/(n*p);
24     return v;
25 }
```
(perm_test.c)

We need to accumulate counts in a histogram with $t!$ elements. We set up the histogram in lines 5, 8 and 10 where fact returns the factorial of its argument. The implementation of fact is left as an exercise for the reader. Line 6 calculates the expected probability for landing in a particular bin, constant for all bins. Lines 7 and 10 set up a buffer to store a set of t consecutive samples from the input file. It is the relative ordering of this buffer that we will be interested in.

The while loop (lines 11–21) fills in the histogram. The for loop (lines 12–16) loads the buffer b with the next set of t samples. Line 18 calculates f for this set of samples and increments the proper histogram bin (line 19) while tracking this set (line 20). Finally, lines 23 and 24 calculate the χ^2 statistic and return it.

The perm function is,

```
 1 int perm(int t, double *b) {
 2     int r=t,f=0,s,i;
 3     double m;
 4
 5     while (r>1) {
 6         m=-1.0;
 7         for(i=0; i<r; i++) {
```

```
 8                    if (b[i]>m) {
 9                        m=b[i];
10                        s=i;
11                    }
12                }
13                f = r*f+s;
14                m = b[s];
15                b[s] = b[r-1];
16                b[r-1] = m;
17                r--;
18            }
19            return f;
20 }
```
(perm_test.c)

which directly implements the f algorithm above. The main `while` loop executes until $r = 0$. At each iteration we will find the largest remaining value (`m`, lines 8–11) and put it into the proper place in the vector `b` (lines 14–16). It is line 13 that we care about. This is the f update and when the loop ends we will return it as the mapping between the initial ordering of the elements and an index into the measurement histogram.

4.1.8 Random Excursions Test

The tests in Sect. 4.1.1 through Sect. 4.1.7 are all based on Knuth [3]. The random excursions test is from the NIST random number test suite [4]. Unlike the tests above that treat the input values as either integers or floating-point values, this test treats the input as a stream of bits. It performs a one-dimensional random walk interpreting each zero bit as a step down (-1) and each one bit as a step up $(+1)$. The test sums the bits across the entire input sequence generating a new sequence of partial sums; each element of the sequence is the sum of all the bits up to that position in the input. The new sequence is then examined for zero crossing points, places where the walk started from zero and has returned to zero. Each zero point marks a cycle. The point of the test is to build a histogram counting the occurrence of each position (deviation) from zero in the range $\{-4, -3, -2, -1, +1, +2, +3, +4\}$. Once we have this histogram we can apply a χ^2 test. We will talk about the expected values for each bin below.

For example, if the input sequence is, bit by bit,

$$1\,0\,1\,1\,1\,0\,0\,0\,1\,0\,1\,1\,1\,1\,1\,0\,0\,1\,0\,1\,0\,1\,0\,1\,1\,0\,0\,0\,0\,0\,1\,1$$

mapping $0 \to -1, 1 \to +1$, gives,

$$1, -1, 1, 1, 1, -1, -1, -1, 1, -1, 1, 1, 1, 1, 1, -1, -1,$$

$$1, -1, 1, -1, 1, -1, 1, 1, -1, -1, -1, -1, -1, 1, 1$$

leading to a cumulative sum sequence of,

$$1, \mathbf{0}, 1, 2, 3, 2, 1, \mathbf{0}, 1, \mathbf{0}, 1, 2, 3, 4, 5, 4, 3, 4, 3, 4, 3, 4, 3, 4, 5, 4, 3, 2, 1, \mathbf{0}, 1, 2$$

where we see there are four zero crossings (marked in bold). The first cycle is of length one and has one +1 step so the histogram value for +1 would be incremented. The second cycle is of length five and has two 1's, two 2's, and one 3 and so on.

In order to apply the χ^2 test we need to know the expected probability of visiting positions from -4 to $+4$. The derivation of these probabilities is outlined in [4] (see Section 3.14, page 3–22, and references therein). The relevant formulas for us are the probabilities associated with the bins of the histogram. These are given as,

$$p_0 = 1 - \frac{1}{2|x|}$$

$$p_k = \frac{1}{4x^2} \left(1 - \frac{1}{2|x|} \right)^{k-1}, \quad k = 1, 2, 3, 4$$

$$p_5 = \frac{1}{2|x|} \left(1 - \frac{1}{2|x|} \right)^4$$

where k is the bin number and x is the label for the bin, $\{-4, -3, -2, -1, +1, +2, +3, +4\}$. Multiplying these probabilities by the number of cycles will give us the expected value for each bin allowing us to perform the χ^2 test using five degrees of freedom for each value of x.

In code, we need a simple driver along with the table of probabilities generated by the equation above. Specifically,

```
 1 #define N 10000000
 2 int8_t B[N];
 3 int16_t S[N];
 4 int T[8][6];
 5 int V[] = {-4,-3,-2,-1,1,2,3,4};
 6 double P[][6] ={{0.8750,0.0156,0.0137,0.0120,0.0105,0.0733},
 7                 {0.8333,0.0278,0.0231,0.0193,0.0161,0.0804},
 8                 {0.7500,0.0625,0.0469,0.0352,0.0264,0.0791},
 9                 {0.5000,0.2500,0.1250,0.0625,0.0312,0.0312},
10                 {0.5000,0.2500,0.1250,0.0625,0.0312,0.0312},
11                 {0.7500,0.0625,0.0469,0.0352,0.0264,0.0791},
12                 {0.8333,0.0278,0.0231,0.0193,0.0161,0.0804},
13                 {0.8750,0.0156,0.0137,0.0120,0.0105,0.0733}};
14
15 int main(int argc, char *argv[]) {
16     buildS(argv[1]);
17     excursions();
18 }
   (excursions_test.c)
```

where N is the number of bits we will work with, the input file must have at least this many bits. Line 2 defines B which stores the actual N bits of the input already converted to -1 or $+1$. Line 3 stores the cumulative sum of these converted bits in S. The histograms (T) are defined in line 4. There are eight possible x values (the values of V, line 5) and each one is a histogram of six values. The fixed bin probabilities are stored in P, line 6, which are directly computed from the equations above. Finally, a very simple driver program first builds the cumulative sums (line 16) and then computes the associated χ^2 and p-values (line 17).

The builds function is divided into three distinct sections. First,

```
1   void buildS(char *src) {
2       int v,i,j,k=0,b,z,w=0;
3       FILE *g;
4
5       g = fopen(src,"r");
6       for(i=0; i<(N/8); i++) {
7           b = next(g);
8           if (b == -1) {
9               printf("The source file must have at least %d
                bits\n",N);
10              exit(0);
11          }
12          B[k++] = (b & 128) ? 1 : -1;
13          B[k++] = (b &  64) ? 1 : -1;
14          B[k++] = (b &  32) ? 1 : -1;
15          B[k++] = (b &  16) ? 1 : -1;
16          B[k++] = (b &   8) ? 1 : -1;
17          B[k++] = (b &   4) ? 1 : -1;
18          B[k++] = (b &   2) ? 1 : -1;
19          B[k++] = (b &   1) ? 1 : -1;
20      }
21      fclose(g);
```
(excursions_test.c)

where the next function in this case simply returns the next byte or -1 if the file is empty. Since N counts bits we need to read N/8 bytes (line 6). If we do get a -1 there are too few bits in the file and the test is abandoned (line 8). Lines 12–19 simply process each bit of the byte storing $+1$ if the bit is 1 or -1 if the bit is 0. The current index is in k.

Once we have B we can build fill in the cumulative random walk vector, S,

```
22      k=0;
23      S[0] = B[k++];
24      for(i=1; i<N; i++) {
25          S[i] = B[k++] + S[i-1];
26      }
```

where each entry in s is the corresponding entry of B plus the previous entry of s (line 25). It is s that we examine for zero crossings. We do this in multiple passes,

```
27      for(j=0; j<8; j++) {
28          v = V[j];
29          for(k=0; k<6; k++) {
30              b=z=0;
31              for(i=0; i<N; i++) {
32                  if (S[i] == v) b++;
33                  if (S[i] == 0) {
34                      if (b==k) z++;
35                      if ((k==5) && (b>5)) z++;
36                      b=0;
37                      w++;
38                  }
39              }
40              T[j][k] = z;
41          }
42      }
43      if (w<500) {
44          printf("Too few cycles, test fails\n");
45          exit(1);
46      }
47  }
```

where j corresponds to x and is the index into V so that v is the current x value (line 28). For the current x we fill in each of the six bins (line 29) with the current bin stored in k. For the current x and k we make a pass through s (line 31). If the current value of s is v it is an instance of the particular value we are interested in at the moment so we count its occurrence (line 32). If the current value of s is zero we have just finished a cycle (line 33). If a cycle is finished we decide if we want to count it. We count it if we counted k occurrences, the current bin of the current x histogram, with a special check for >5 if $k = 5$ (line 35). We then reset b (line 36) since the cycle has ended and increment w which counts the number of zero crossings found. Finally, we are able to fill in the k-th entry of the x-th histogram in line 40.

We repeat the above for all eight histograms and all six bins per histogram to completely fill in the T table. There is then a final check on w to ensure that a minimum number of zero crossing were found (line 43). If we find less than 500 in 10,000,000 bits the test results, based on the same theory that led to the bin probabilities, are not to be considered valid and the test exits with a warning.

With T filled in we can perform the eight χ^2 tests using excursions,

```
 1  void excursions(void) {
 2       double n,v,p;
 3       int i,j;
 4
 5       for(i=0; i<8; i++) {
 6           v = 0.0;
 7           n=0;
 8           for(j=0; j<6; j++)
 9               n += T[i][j];
10           for(j=0; j<6; j++) {
11               v += (T[i][j]-n*P[i][j])*(T[i][j]-n*P[i][j])/
                     (n*P[i][j]);
12           }
13           p = chisq_pval(v,5);
14           printf("v=%2d, chisq=%0.4f, p=%0.7f\n",V[i],v,p);
15       }
16  }
```
(excursions_test.c)

where for each of the eight x values (i) we determine the sum across the histogram (n, lines 7–9) and use it to calculate the χ^2 statistic (v, lines 10–12). This value, and its p-value (line 13), are then reported.

The random excursions test is passed if, for each of the possible x values, the p-value is >0.05 and <0.95, as before.

4.2   Applying the Classical Randomness Tests

The preceding sections have defined several randomness tests. Let's now apply them to the generators we developed in Chap. 2 along with data we know (strongly believe) to be truly random. For the latter, we will use a large collection of random bits courtesy of Mads Haahr from RANDOM.ORG [5]. These bits are derived from atmospheric noise; see the website for details. We will also use the TV image data set from Chap. 1.

Let's start with the smaller TV data set. We can run the Knuth-derived tests on double precision floats made by grouping the input into unsigned 32-bit integers divided by 2^{32}. We will use the raw bits for the random excursions test. For example, the χ^2 test is run with,

```
./chisq 1 random_tv.dat
```

where the "1" interprets the input file as a series of unsigned 32-bit integers that are converted to double precision floats. The output is,

```
chisq = 4497.62992409, p-value = -nan
```

indicating that the χ^2 test has failed with a p-value that is a NaN (not a number). Not a good start. Let's see what the other Knuth tests have to say about the TV data.
Repeating this process for the remaining Knuth-derived tests gives,

```
KS-test        : kp=0.758947, km=0.252982
Serial         : chisq=2211.00966, p-value=-nan
Permutation(3): chisq=3.30926057, p-value=0.23085088
Max-of-5       : chisq=448.64495421, p-value=0.00000000
Correlation    : corr=0.00014, 95% CI=[-0.00076, 0.00076]
```

where we see that the KS-test is passed because both the largest positive (kp) and negative (km) deviations are within the $0.1548 < K_n^+, K_n^- < 1.2186$ range. The permutation test has passed with a p-value of 0.23. The correlation test has also passed. However, the serial test and max test have both failed. So, we have three tests that passed and three tests that have failed.

Observant readers will have noticed that we ignored the gap tests. This is not an oversight but intentional. The gap test requires at least 100 million samples or more to produce meaningful results. To apply it to a file as small as the TV data would be misleading.

Let's run the NIST random excursions test. Doing so gives,

```
v=-4, chisq=4.1050, p=0.4655968
v=-3, chisq=4.2589, p=0.4872200
v=-2, chisq=3.2624, p=0.3403998
v=-1, chisq=6.1226, p=0.7055272
v= 1, chisq=1.3712, p=0.0725673
v= 2, chisq=1.4861, p=0.0853292
v= 3, chisq=0.4591, p=0.0064541
v= 4, chisq=8.9185, p=0.8876409
```

which shows that the χ^2 tests for all positions except v=3 are random. The v=3 test is below our conservative threshold but not dramatically so.

While repeated tests are not trivial for the TV data, it is trivial when we decide to test the pseudorandom generators of Chap. 2. There is no reason why we cannot create many dozens of Mersenne Twister output files, each with a different starting seed value, and then test them all to see if there are any trends that indicate weakness in the algorithm. Will all tests be passed at all times? Probably not. This is a stochastic process so even a truly random process will, from time to time, create an output sequence that fails a particular test.

Let's run the experiment. Using a series of Python scripts (see the website for this book), we can run our randomness tests on each combination of test and pseudorandom generator using 50 different outputs from each generator. This means we will have 50 separate χ^2 test results for the Mersenne Twister generator, as well as all other generators, and from this we can count the number of times the test fails. This will give us valid statistics on how often generators fail particular tests but we need something to compare the results with. This is where the RANDOM.ORG bits come in. We can use 50 sets of bits in the same way that we use the 50 pseudorandom outputs. This will give us an idea of how a truly random sequence behaves in regards

Table 4.1 Fraction of test results failing the particular test. Code for these pseudorandom generators is in Chap. 2 except for the Middle Weyl generator which is in Chap. 1

Generator	χ^2	KS	Serial	Permutation	Maximum	Correlation	Gap (below, above)
RANDOM.ORG	0.08	0.00	0.12	0.20	0.06	0.00	n/a
Mersenne Twister	0.06	0.10	0.16	0.18	0.18	0.08	(0.26, 0.14)
xorshift32	0.06	0.08	0.12	0.14	0.06	0.00	(0.12, 0.20)
xorshift1024*	0.08	0.04	0.10	0.12	0.10	0.06	(0.20, 0.22)
CMWC	0.08	0.02	0.10	0.12	0.10	0.12	(0.20, 0.12)
KISS	0.16	0.16	0.08	0.06	0.14	0.06	(0.30, 0.12)
Middle Weyl	0.10	0.02	0.04	0.12	0.06	0.06	(0.30, 0.34)
MINSTD (original)	0.20	0.04	0.12	0.28	0.08	0.02	(0.18, 0.20)
MINSTD (revised)	0.10	0.02	0.12	0.24	0.14	0.06	(0.18, 0.20)
RANDU	0.52	0.06	0.50	0.14	1.00	0.00	(1.00, 1.00)

There was not enough data to run the gap test on sets of RANDOM.ORG bits. When run on the single data file it passed with p-values (0.1783, 0.5645)

to these tests and something against which we can compare the pseudorandom generators. Why 50? No particular reason other than that it is a good compromise between having enough data for a meaningful result and the amount of time it takes to generate and test files of random values. We will see in Sect. 4.3 a more rigorous way to interpret the output of the randomness tests but for now let's proceed with our ad hoc approach.

Specifically, then, we will use the following tests, each with 50 million double precision values from the given generators, each with a different 32-bit seed: χ^2, Kolmogorov-Smirnov, serial, permutation, maximum, gap, and correlation. For the gap test we will use 150 million values to gain valid statistics. For the permutation test we will use $t = 3$. For the maximum-of-t test we will use $t = 5$. The specific test results are on the book website. What we will look at here is the fraction of tests that failed by producing results that were out of our conservative range. For tests producing a χ^2 value this means p-values <0.05 or >0.95. For the KS-test it means deviations outside $0.1548 < K_n^+, K_n^- < 1.2186$. Finally, for the correlation test, we look for correlations outside of the 95% bounds for the number of samples. All of these tests, except the gap test, are repeated for the 50 RANDOM.ORG files interpreted as double precision floats. The results are in Table 4.1.

Table 4.1 shows clearly that even a truly random sequence will fail these tests from time to time. This is expected. The fraction of failures is our guide for how often we would expect a good pseudorandom generator to fail the same tests. For example, if we look at the absolute deviation, per test, between the RANDOM.ORG data and the Mersenne Twister we get a vector,

$$(0.02, 0.10, 0.04, 0.02, 0.12, 0.08)$$

with norm 0.1822 (we ignored the gap test results). Doing the same with the RANDU generator gives a vector of (0.44, 0.06, 0.38, 0.06, 0.940.00) with norm

1.1085. Clearly, the RANDU generator is in some sense "further" from the RAN-DOM.ORG data than the Mersenne Twister is. Using the norm of this vector as a ranking metric we can score the generators and arrange them from "best" to "worst",

Generator	Norm
xorshift32	0.1020
MINSTD (revised)	0.1114
xorshift1024*	0.1166
Middle Weyl	0.1311
CMWC	0.1523
MINSTD (original)	0.1523
Mersenne Twister	0.1822
KISS	0.2514
RANDU	1.1085

By this ad hoc ordering we see that the xorshift32 generator has test results most similar to the RANDOM.ORG bits. The next six generators can be viewed as a group. The KISS generator is a little further out. And, as would be expected, the terrible RANDU generator is over four times as far from the RANDOM.ORG bits than even the KISS generator. Graphically, the distance between RANDU and the other generators is even more striking as Fig. 4.1 shows.

To add further perspective, let's see how the RANDOM.ORG data compares to a second source of truly random bits, those of the HotBits web page [6]. The bits in these data files are generated from the decay of a radioactive isotope. In this case, we downloaded the 11 MB test file and separated it into 20 smaller files, each with 143,360 double precision floats. The floats were calculated by grouping the bytes into sets of four and dividing by 2^{32}, exactly the same approach used for the RANDOM.ORG data. If we calculate the fraction of the six tests which fail over these 20 files we get a vector,

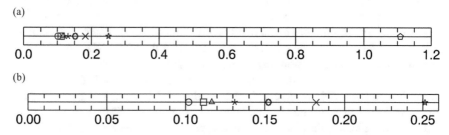

Fig. 4.1 A graphical representation of the ordering of the tested pseudorandom generators. The plot shows the norm of the vector difference between the given generator and the truly random values from RANDOM.ORG. The smaller the norm the more similar the generator is to the test results of the truly random data. In (**a**) all generators are shown with RANDU at the far right (pentagon). In (**b**) the remaining generators are separated by reducing the range of the x-axis: Middle Weyl (circle), xorshift1024* (square), xorshift32 (triangle), original MINSTD (asterisk), CMWC (diamond), revised MINSTD (hexagon), Mersenne Twister (X), and KISS (star)

$$(0.15, 0.05, 0.10, 0.10, 0.00, 0.00)$$

which we can compare to RANDOM.ORG data to get a difference vector and norm. The norm is 0.1463 which places the HotBit data between the Middle Weyl and CMWC generators. This is interesting and telling. It means that essentially, all the generators tested are really no different, according to these tests and metric, than one source of truly random data is from another source of truly random data. The clear exception is the RANDU generator.

Ignoring RANDU, does this mean that the generators tested are interchangeable? No. It does mean that, as far as these tests can discern, for double precision floats produced by these generators, the output of one is as "random" as another. This does not mean that the generators are all equally useful. If we need to generate billions upon billions of samples, clearly we should not consider the MINSTD linear congruential generator with its short period. Instead, we should look to the generators with long periods like the Mersenne Twister ($2^{19,937} - 1$), CMWC ($2^{131,086}$), or the insane period of the KISS generator ($2^{40,000,000}$). On the other hand, if we are in an embedded or small system environment we should consider simple to implement generators like the xorshift family. As we saw in Sect. 2.4, even a simple 8-bit microprocessor can support the 32-bit version of xorshift.

Naturally, the classical tests presented here are not the final word on the matter. In Chap. 2 we saw that many of these generators have specific issues such as equidistribution of bits or small periods that make them unsuitable for certain tasks. Since it is impossible to clearly state that a sequence of values is, definitively, random, we must rely on accumulated evidence based on different tests one might apply to such a sequence. The classical tests here (lumping the newer NIST random excursions test in for good measure) are able to hint at randomness but researchers were dissatisfied with only these tests and sought for a suite of tests that might be used by the community to rank and compare generators with some measure of confidence and consistency. We will examine two of these test suites in the following sections and in the process re-evaluate some of the generators of this section.

4.3   Test Suite—Dieharder

The dieharder random number test suite of Brown [1], built on the older DIEHARD test suite of Marsaglia [7], is in active development at the time of this writing and includes tests from DIEHARD, rewritten in C, as well as some from the NIST random number suite, along with original tests from Brown and several others. The source package,

http://webhome.phy.duke.edu/~rgb/General/dieharder.php

will compile on standard Linux systems out of the box with the usual Unix sequence,

```
$ ./configure
$ make
$ sudo make install
```

and once built and installed is run from the command line,

```
$ ./dieharder
```

To see the list of available generators, those that are built into the application, use,

```
$ ./dieharder -g -1
```

and to see the list of available tests,

```
$ ./dieharder -l
```

These are based on older DIEHARD tests and include NIST tests (marked "STS") as well as versions from Brown (marked "RGB") and a few others.

The dieharder program is quite configurable and is meant for extreme testing of generators. A full exploration of its capabilities is well beyond the introduction we can give here. However, to simply run the tests with default parameters one need only use the -a option. The tests themselves are described with,

```
$ ./dieharder -a -h
```

Just as we did for the Knuth tests, dieharder runs each test many times to generate a set of output p-values instead of a single output value. See the dieharder man page for an excellent account of this process. The output report includes a final p-value along with the program's assessment of whether or not the test could be considered to have passed. Recall that this is necessary because, as we saw in Sect. 4.1, even a purely random sequence (or, what we are here calling purely random) will fail the tests from time to time, so a single run is not sufficient to decide success or failure.

We will start with some of the built-in generators that are familiar to us: MIN-STD, Mersenne Twister, KISS, L'Ecuyer's combined generator, and, for illustration purposes, RANDU. To run the tests with the MINSTD generator use,

```
$ ./dieharder -a -g 11
```

which produces a report starting with,

```
#=============================================================================#
#            dieharder version 3.31.1 Copyright 2003 Robert G. Brown      #
#=============================================================================#
   rng_name    |rands/second|   Seed      |
        minstd|   1.13e+08  |1419740157|
#=============================================================================#
         test_name   |ntup| tsamples |psamples|   p-value  |Assessment
#=============================================================================#
   diehard_birthdays|   0|      100|  100|0.56250254|   PASSED
      diehard_operm5|   0|  1000000|  100|0.29594541|   PASSED
  diehard_rank_32x32|   0|    40000|  100|0.20679229|   PASSED
    diehard_rank_6x8|   0|   100000|  100|0.71945062|   PASSED
   diehard_bitstream|   0|  2097152|  100|0.83939659|   PASSED
        diehard_opso|   0|  2097152|  100|0.00151752|     WEAK
        diehard_oqso|   0|  2097152|  100|0.00041954|     WEAK
         diehard_dna|   0|  2097152|  100|0.00000000|   FAILED
 diehard_count_1s_str|   0|   256000|  100|0.58543810|   PASSED
```

showing the program version, the generator being tested, how quickly the test machine is able to create samples from the generator, and the starting seed value. Next comes a long list of test results in six columns including the test name, the number of samples used by the test (`tsamples`), the number of p-values generated (`psamples`) which is the number of times the test is repeated, an overall p-value, and dieharder's assessment of the results showing if the test should be considered to have PASSED, FAILED, or has given WEAK results, similar to Knuth's "suspicious" range.

The previous paragraph slipped something in, the "overall p-value". What does that mean? If we repeat a test returning a p-value we should, for different starting seeds, expect the distribution of p-values to be uniform over [0, 1]. We know how to test if a set is uniformly distributed from the tests of Sect. 4.1. We can run a χ^2 test or a Kolmogorov-Smirnov test (KS test). The dieharder program uses the KS test on the distribution of p-values to determine the final p-value reported. Further, the final assessment is dieharder applying its own internal p-value thresholds to return PASSED, WEAK, or FAILED. These criteria are,

FAILED	$p < 0.0005$ or $p > 0.9995$
WEAK	$p < 0.005$ or $p > 0.995$
PASSED	any other p-value

We can make dieharder show us, for a specific test, the distribution of p-values it found. For example, to run the 2D sphere test (try `dieharder -d 11 -h`) on the MINSTD generator visualizing the histogram of p-values enter,

```
dieharder -d 11 -g 11 -D default -D histogram
```

where `-d 11` selects the 2D sphere test and `-g 11` selects the MINSTD generator. Your result will look different but it will be similar to this,

```
#=======================================================================#
#                       Histogram of test p-values                      #
#=======================================================================#
# Bin scale = 0.100000
#    20|      |      |      |      |      |      |      |      |      |
#      |      |      | **** |      |      |      |      |      |      |
#    18|      |      | **** |      |      |      |      |      |      |
#      |      |      | **** |      |      |      |      |      |      |
#    16|      |      | **** |      |      |      |      |      |      |
#      |      |      | **** |      |      |      |      |      |      |
#    14|      |      | **** |      |      |      |      |      |      |
#      | **** |      | **** |      |      |      | **** |      |      |
#    12| **** |      | **** |      |      |      | **** |      |      |
#      | **** |      | **** |      | **** | **** | **** |      |      |
#    10| **** |      | **** |      | **** | **** | **** |      |      |
#      | **** |      | **** | **** | **** | **** | **** |      |      |
#     8| **** | **** | **** | **** | **** | **** | **** |      |      |
#      | **** | **** | **** | **** | **** | **** | **** | **** | **** |
```

```
#    6|****|****|    |****|****|****|    |****|****|****|
#     |****|****|    |****|****|****|****|****|****|****|
#    4|****|****|****|****|****|****|****|****|****|****|
#     |****|****|****|****|****|****|****|****|****|****|
#    2|****|****|****|****|****|****|****|****|****|****|
#     |****|****|****|****|****|****|****|****|****|****|
#     |--------------------------------------------------
#     | 0.1| 0.2| 0.3| 0.4| 0.5| 0.6| 0.7| 0.8| 0.9| 1.0|
#=====================================================================#
#=====================================================================#
         test_name    |ntup| tsamples |psamples|  p-value |Assessment
#=====================================================================#
      diehard_2dsphere|   2|     8000|      100|0.39255511|   PASSED
```

showing that MINSTD passed the 2D sphere test with a KS test *p*-value, applied to the 100 *p*-values in the histogram, of 0.3926, well within the limits.

If we apply the same test to the RANDU generator (`-g 41`) we get something like this,

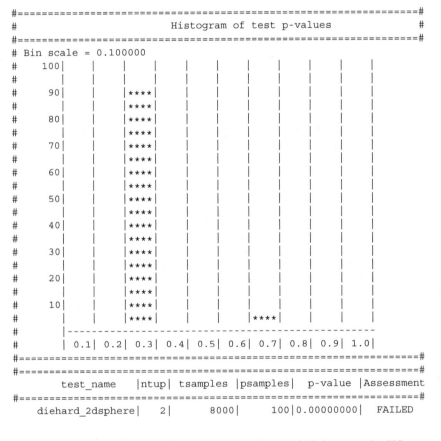

```
#=====================================================================#
#                     Histogram of test p-values                     #
#=====================================================================#
# Bin scale = 0.100000
#    100|    |    |    |    |    |    |    |    |    |    |
#       |    |    |    |    |    |    |    |    |    |    |
#    90 |    |    |****|    |    |    |    |    |    |    |
#       |    |    |****|    |    |    |    |    |    |    |
#    80 |    |    |****|    |    |    |    |    |    |    |
#       |    |    |****|    |    |    |    |    |    |    |
#    70 |    |    |****|    |    |    |    |    |    |    |
#       |    |    |****|    |    |    |    |    |    |    |
#    60 |    |    |****|    |    |    |    |    |    |    |
#       |    |    |****|    |    |    |    |    |    |    |
#    50 |    |    |****|    |    |    |    |    |    |    |
#       |    |    |****|    |    |    |    |    |    |    |
#    40 |    |    |****|    |    |    |    |    |    |    |
#       |    |    |****|    |    |    |    |    |    |    |
#    30 |    |    |****|    |    |    |    |    |    |    |
#       |    |    |****|    |    |    |    |    |    |    |
#    20 |    |    |****|    |    |    |    |    |    |    |
#       |    |    |****|    |    |    |    |    |    |    |
#    10 |    |    |****|    |    |    |    |    |    |    |
#       |    |    |****|    |    |    |****|    |    |    |
#       |--------------------------------------------------
#       | 0.1| 0.2| 0.3| 0.4| 0.5| 0.6| 0.7| 0.8| 0.9| 1.0|
#=====================================================================#
#=====================================================================#
         test_name    |ntup| tsamples |psamples|  p-value |Assessment
#=====================================================================#
      diehard_2dsphere|   2|     8000|      100|0.00000000|   FAILED
```

showing yet once again how terrible RANDU is. The test fails because the KS test *p*-value is below 0.0005.

Let's evaluate several built-in generators using the default tests and options (-a option). The dieharder default test suite conducts 114 tests, many are the same test applied to different bits of the output, and for each test dieharder reports PASSED, WEAK, or FAILED. One way people summarize generator test results is by stating the number of tests that the generator has passed. For the built-in generators we get the following,

	Passed	Weak	Failed
MINSTD	104	4	6
Mersenne Twister	111	3	0
KISS	113	1	0
Combined L'Ecuyer	111	2	1
RANDU	36	5	73

which we can summarize still further by applying a simple equation, $s = 2p + w - 2f$, to give a single score enabling ranking of the results. Naturally, there are many such equations we could use, the chosen one gives double credit to a passing test, single credit to a weak test, and double negative credit for a failure. A generator passing all tests will have a score of 228. A generator failing all tests will have a score of -228. Scaling by 228 would map these into a $[-1, +1]$ range, if desired.

Scoring the generators gives a ranking of,

	Score
KISS	227
Mersenne Twister	225
Combined L'Ecuyer	222
MINSTD	200
RANDU	−69

which tracks nicely with how we have been interpreting the quality of the various generators, especially as they were defined in Chap. 2. Our ad hoc distance from the results of the Knuth tests on truly random bits from Sect. 4.2 are in general agreement, though that metric placed MINSTD closer to random data than it really should have been. Still, both metrics show that RANDU is a truly terrible pseudorandom number generator.

What if dieharder does not natively support the generator we wish to test? One option is to generate a large file of samples from the generator and then pass them to dieharder for testing. Let's look at how to do that for the xorshift and CMWC generators as well as the RANDON.ORG, TV and HotBits files.

First, we need a file of samples. In Sect. 4.2 we used double precision floating-point samples from the generators, here we instead use files of integer samples which dieharder will read in blocks of 32 bits. For this we use a modified version of the ugen program from Sect. 3.1. This program is available on the book website. With this program we created files with 500 million samples each (2 billion bytes) for Marsaglia's original xorshift32 and xorshift128 generators. We also created files for the xorshift128+ and xorshift1024* variants. Finally we created a 500 million sample file for the CMWC generator. See Sects. 2.4 and 2.5 for definitions of these generators.

Each file was passed to dieharder's default tests,

```
$ dieharder -a -g 201 -f tmp.dat
```

with tmp.dat the file containing the samples. If, during a test, dieharder runs out of samples it rewinds the file and reads again. With 500 million samples per file, dieharder may run out of samples but this is okay because we will still get valid statistics. With smaller files the results will be unreliable.

Repeating the default tests as we did for the built-in generators gives the following results,

	Passed	Weak	Failed
xorshift32	107	3	4
xorshift128	109	5	0
xorshift128+	111	3	0
xorshift1024*	112	2	0
CMWC	112	2	0

which we can score to rank the inputs giving,

	Score
CMWC	226
xorshift1024*	226
xorshift128+	225
xorshift128	223
xorshift32	209

showing that the xorshift family, along with the CMWC generator, are also quite good, as we already suspected.

To this point we have ignored the Middle Weyl generator even though it (seemingly) did well with the classical randomness tests of Sect. 4.1. This was intentional. True, we could generate a large file of samples as we did for the xorshift family of generators but instead we will enhance dieharder by adding it to the built-in generators.

The dieharder source code includes a sample generator file that we can use as a template for the Middle Weyl generator: rng_empty_random.c. Following this file, we create a new file, rng_middle_weyl.c containing,

```
 1 #include "dieharder.h"
 2 #include <stdint.h>
 3 uint64_t mw_seed = 0xb5ad4ece00000000;
 4
 5 unsigned long int mw_random(void) {
 6     static uint64_t x=0, w=0;
 7     x *= x;
 8     w += mw_seed;
 9     x += w;
10     x = (x >> 32) | (x << 32);
11     return (unsigned long int)x;
12 }
13
14 static unsigned long int mw_random_get (void *vstate) {
15     return mw_random();
16 }
17
18 static double mw_random_get_double (void *vstate) {
19     return (double)mw_random() / 4294967296.0;
20 }
21
22 static void mw_random_set (void *vstate, unsigned long {
   int s) {
23     mw_seed |= (uint64_t)seed;
24 }
25
26 typedef struct {} mw_random_state_t;
27 static const gsl_rng_type mw_random_type =
28 {"middle_weyl",          /* name */
29  4294967295UL,  /* RAND_MAX */
30  0,             /* RAND_MIN */
31  sizeof (mw_random_state_t),
32  &mw_random_set,
33  &mw_random_get,
34  &mw_random_get_double};
35 const gsl_rng_type *gsl_rng_mw_random = &mw_random_type;
```
(rng_middle_weyl.c)

where lines 3–12 implement the generator. Lines 14–24 define an interface used by dieharder to get samples from the generator as 32-bit unsigned integers or double precision floats. The mw_random_set function sets the seed by OR-ing in the unsigned 32-bit seed supplied by dieharder to the 64-bit unsigned integer seed the Middle Weyl generator requires. This explains the lower 32-bits of line 3 being zero. Finally, lines 26–35 define more expected interface code following the GSL (Gnu Scientific Library) format so that dieharder will run.

There are several other steps necessary to get dieharder to recognize and use the new generator. First, we need to add the new generator to dieharder.h like so,

```
GSL_VAR const gsl_rng_type *gsl_rng_empty_random;
GSL_VAR const gsl_rng_type *gsl_rng_mw_random;
```

and then to add_ui_rngs.c,

```
i = 600;
dh_num_user_rngs = 0;
dh_rng_types[i] = gsl_rng_mw_random;
i++;
dh_num_user_rngs++;
dh_num_rngs++;
```

followed by ensuring that dieharder is aware of the new generator by adding a call to add_ui_rngs to dieharder.c after it initializes the library generators,

```
dieharder_rng_types();
add_ui_rngs();
```

At this point the code has been modified. Now we need to update the build environment by adding rng_middle_weyl.c to Makefile.am under dieharder_SOURCES,

```
dieharder_SOURCES = \
    add_ui_rngs.c \
    add_ui_tests.c \
    choose_rng.c \
    dieharder.c \
    dieharder_exit.c \
    help.c \
    list_rngs.c \
    list_tests.c \
    output.c output.h \
    output_rnds.c \
    parsecl.c \
    rdieharder.c \
    run_all_tests.c \
    run_test.c \
    set_globals.c \
    testbits.c \
    time_rng.c \
    user_template.c \
    rng_middle_weyl.c
```

and finally we are ready to reconstruct dieharder,

```
autoreconf
./configure
make
sudo make install
```

To check that the new generator is listed use,

```
$ ./dieharder -g -1
```

to see that generator 600 is now present and called `middle_weyl`. Run the default tests with,

```
$ ./dieharder -a -g 600
```

The default tests give excellent results: 112 PASSED, 2 WEAK, and 0 FAILED, for a score of 226. Adding these to our list of tested generators, we get a final list of scores, from best to worst,

	Score
KISS	227
Middle Weyl	226
xorshift1024*	226
CMWC	226
Mersenne Twister	225
xorshift128+	225
xorshift128	223
Combined L'Ecuyer	222
xorshift32	209
MINSTD	200
RANDU	-69

which tracks the intuition we have developed in this chapter and in Chap. 2 where we implemented the generators. For example, the xorshift family falls out exactly as we might expect with xorshift32 the least effective to xorshift1024* as the most effective.

We also ran the default diehard tests against the files of random samples from RANDOM.ORG, HotBits, and our TV data set from Chap. 1 with the following results,

	Passed	Weak	Failed	Score
RANDOM.ORG	110	3	1	221
HotBits	53	14	47	26
TV	26	5	83	−109

which seems to imply that the HotBits dataset is not particularly random. In fact, it likely is, the HotBits site has its own tests on larger files, but we failed here because the number of available samples is too low. Recall that dieharder processes input files in 32-bit chunks. The HotBits dataset has only 2,867,200 samples. However, the RANDOM.ORG data set, which gave good results, has 101,974,016 samples.

Therefore, if we were to generate at least as many samples from HotBits we should (careful with "should") get results similar to the RANDOM.ORG sample. Let this be a cautionary tale about attempting to decide the randomness of a small file using powerful random number test suites. We will see in Sect. 4.5 an approach we can use to gain insight into the randomness of smaller files.

What about the TV data set? Clearly, something is amiss. Some of the failures are probably due to the small file size, just under 7 million 32-bit samples, but that likely does not account for all the failures. The TV dataset was generated from digital photos of an actual CRT television tuned to an empty channel. While the underlying noise signal shown by the TV is highly random, the *process* of turning the image into a stream of bytes has clearly introduced bias. The images were 1500× 1500 pixels and resampled, with nearest neighbor sampling to not introduce artificial values, to 500 × 500 pixels before processing. Even at that resolution it is likely that any selected pixel will have neighbors that are very much like it. We attempted to get around this somewhat by pulling samples from other areas of the image, to incorporate a spatial component, but clearly the approach is not as tuned as it could be and a bias exists which leads to poor performance on statistical tests. Recall, also, that this dataset failed half of the classical randomness tests of Sect. 4.1. This is a dramatic example but subtle biases have been found in other, more carefully tuned, true random number generators.

In this section we saw how to install and use, at a very basic level, the dieharder test program. In truth, we have barely scratched the surface of what dieharder can really do and how it is really meant to be used. It is a program for statisticians and mathematicians who wish to strenuously test the limits of pseudorandom number generators. We have used it here as a tool to help us understand how to decide if we should use a generator and to deliver some metrics on how "good" a generator might actually be.

We saw, especially when compared to truly random data (?) like the bits from RANDOM.ORG, that a number of the generators from Chap. 2 are really quite good. The enhanced xorshift generators, the KISS generator, the Mersenne Twister, the CMWC generator and the new Middle Weyl generator all gave excellent results. When to use any of these depends upon the actual application. A Monte Carlo system might want the long period of KISS, CMWC, or the Mersenne Twister. An embedded system might be able to use xorshift128+ or even xorshift32 for the simplest of processors. As expected, the MINSTD generator, being a single linear congruential generator, was weaker than the more advanced generators but is still, perhaps for a simple game, more than adequate. Avoid RANDU at all costs.

4.4 Test Suite—TestU01

The TestU01 suite [2] is a C software library for testing pseudorandom generators. Like dieharder (Sect. 4.3) it contains a number of predefined generators as well as an extensive library of tests. The tests are separated into three groups which take correspondingly longer to execute. The smallest is the *SmallCrush* suite which executes in a few seconds. This is followed by the *Crush* suite (tens of minutes) and the *BigCrush* suite which takes hours to complete.

Building TestU01 is straightforward. Download the TestU01 sources from,

http://simul.iro.umontreal.ca/testu01/tu01.html

unzip and build with the usual sequence,

```
$ ./configure
$ make
$ sudo make install
```

which will build the libraries and install them in /usr/local/ directories. Specifically, it will create,

```
/usr/local/lib/libtestu01.so
                libprobdist.so
                libmylib.so
```

along with a few dozen include files in /usr/local/include.

Rather than repeat the tests we performed in Sect. 4.3 using the TestU01 prebuilt implementations, we will instead implement two simple programs. The first will again test the Middle Weyl generator, which is not part of the TestU01 suite, and the second will test a file of unsigned 32-bit integers or double precision floats.

The TestU01 library uses the concept of a "generator" object which is created, along with any necessary parameters, and then passed to the batteries of tests. Both of our sample programs will make use of the unif01_CreateExternGenBits function to create a generator that returns 32-bit unsigned integers. Once we have this generator object we can trivially apply the test suites and capture the output for analysis.

With this interface the Middle Weyl generator is straightforward to implement,

```
 1  uint64_t seed = 0xb5ad4ece00000000;
 2
 3  unsigned int middle_weyl(void) {
 4      static uint64_t x=0, w=0;
 5      x *= x;
 6      w += seed;
 7      x += w;
 8      x = (x >> 32) | (x << 32);
 9      return (unsigned int)x;
10  }
11
12  int main(int argc, char *argv[]) {
13      unif01_Gen *gen;
14
15      seed |= atol(argv[1]);
16      gen = unif01_CreateExternGenBits("Middle Weyl",
        middle_weyl);
17
```

```
18      switch (atoi(argv[2])) {
19          case 0:   bbattery_SmallCrush(gen);   break;
20          case 1:   bbattery_Crush(gen);        break;
21          default:  bbattery_BigCrush(gen);     break;
22      }
23
24      unif01_DeleteExternGenBits(gen);
25      return 0;
26  }
```
(testu01_middle_weyl.c)

where lines 1–10 are the Middle Weyl generator from Chap. 1. This generator uses a 64-bit seed value (line 1) where for simplicity we are fixing the upper 32-bits, which cannot be zero, and OR-ing in the lower 32-bits read from the command line (line 15). The seed supplied must be an odd number. Line 16 creates the TestU01 generator object. It accepts a string which is shown in the test output and a function that returns an unsigned integer and accepts no arguments. Lines 18–22 simply select which battery of tests to run. These tests output to stdout so capturing the output of the program will capture the results. Line 24 frees memory used by the generator object.

This program can be trivially modified to work with a generator that returns a double precision float in the range [0, 1). Simply replace unif01_CreateExternGenBit with unif01_CreateExternGen01 and pass in a function returning a double precision float.

Virtually identical programs can be made using the counter-based generator code from Sect. 2.6 so we can test these generators as well. Note, TestU01 expects an unsigned 32-bit integer so the 64-bit output of the Threefry generator is shifted down 32-bits before being returned. It is hoped that this bit of a cheat will be excused. Look for *testu01_threefry.c* and *testu01_philox.c* on the website for this book.

Our second program reads bytes from a file in 32-bit chunks and passes them to TestU01. It is quite similar to the Middle Weyl program but replaces the pseudorandom generator function with a function that reads values from a disk file rewinding the file when it is exhausted. In code,

```
1  FILE *g;
2  int rewinds = 0;
3
4  unsigned int fromfile(void) {
5      unsigned int u32;
6      if (fread((void*)&u32, sizeof(unsigned int), 1, g)
               == 0) {
7          rewind(g);
8          rewinds++;
9          fread((void*)&u32, sizeof(unsigned int), 1, g);
10     }
11     return u32;
12  }
13
```

```
14  int main(int argc, char *argv[]) {
15      unif01_Gen *gen;
16
17      gen = unif01_CreateExternGenBits("From file",
        fromfile);
18      g = fopen(argv[2],"r");
19
20      switch (atoi(argv[1])) {
21          case 0:   bbattery_SmallCrush(gen);   break;
22          case 1:   bbattery_Crush(gen);        break;
23          default:  bbattery_BigCrush(gen);     break;
24      }
25
26      printf("\n\ninput file rewound %d times\n\n",
        rewinds);
27      unif01_DeleteExternGenBits(gen);
28      fclose(g);
29      return 0;
30  }
```
(testu01_from_file.c)

where `fromfile` reads unsigned 32-bit chunks of the given file (line 18) rewinding when necessary. The number of times the file is rewound during the tests is counted in `rewinds` and reported when the tests are complete (line 26). Ideally, the file will have enough values so that it need not be rewound at all.

Both of these programs need to be compiled so that the TestU01 libraries and include files can be found. Use the following,

```
$ gcc from_file.c -o from_file -O3 -Iinclude -Llib -ltestu01
  -lprobdist -lmylib -lm
```

Let's start with the Middle Weyl generator. The SmallCrush suite of 15 tests takes only a few seconds to run. A lot of output is generated and a review of the TestU01 manual will explain it. For us, the relevant section is at the very end,

```
========= Summary results of SmallCrush =========

 Version:           TestU01 1.2.3
 Generator:         Middle Weyl
 Number of statistics:  15
 Total CPU time:    00:00:11.00

 All tests were passed
```

where we see that the Middle Weyl generator passed all the SmallCrush tests. Looking at the Crush output gives a summary of,

```
========= Summary results of Crush =========

 Version:           TestU01 1.2.3
 Generator:         Middle Weyl
```

```
Number of statistics:  144
Total CPU time:   00:38:01.94
The following tests gave p-values outside [0.001, 0.9990]:
(eps  means a value < 1.0e-300):
(eps1 means a value < 1.0e-15):

    Test                              p-value
---------------------------------------------------
22  ClosePairsBitMatch, t = 4        2.2e-4
77  LongestHeadRun, r = 20           0.9992
---------------------------------------------------
All other tests were passed
```

where we see that Middle Weyl failed two of the 144 tests, though it only failed one of them by 0.0002. The Crush suite took 38 min to run.

Finally, we ran BigCrush on Middle Weyl. BigCrush is perhaps the strictest of all pseudorandom number test suites and consists of 160 tests. Middle Weyl's summary is,

```
========= Summary results of BigCrush =========

Version:          TestU01 1.2.3
Generator:        Middle Weyl
Number of statistics:  160
Total CPU time:   04:35:04.30

All tests were passed
```

which is very strong evidence for the quality of this simple generator. Note also that BigCrush took 4.5 h to run.

For comparison, using Threefry passes all Crush tests and all but three Big Crush tests. Philox passes all Big Crush tests and all but one Crush test.

None of the data files we have been using from truly random sources is large enough to run through TestU01 except the RANDOM.ORG data. Running SmallCrush returns,

```
========= Summary results of SmallCrush =========

Version:          TestU01 1.2.3
Generator:        From file
Number of statistics:  15
Total CPU time:   00:00:18.05

All tests were passed

input file rewound two times
```

telling us that all tests were passed but also that the input file, 389 MB in size, was rewound two times during even the SmallCrush tests. TestU01 needs a lot of input data, potentially gigabytes of input data, so it is best used with a software generator or a live hardware stream and not a fixed length file of bytes.

TestU01 is far more than what we have presented here. Like dieharder, TestU01 supplies an extensive set of tools for serious research into pseudorandom number generation. The interested reader is directed to the substantial documentation found in,

```
/usr/local/share/TestU01/doc/
```

where one will find multiple versions of the user manual.

4.5   Quick Randomness Tests—ent

Sometimes, as we saw in the previous two sections, we may have a file of bytes that is too small for the exhaustive tests of dieharder or TestU01. In these situations we can make use of the simpler *ent* program which is included with most common Linux distributions. Unlike the bigger test suites, ent runs a small set of tests and leaves it up to the user to decide if the results of those tests are reasonable or not.

Let's look at some example output to see what ent actually does. Running it on the TV data set,

```
$ ent random_tv.dat
```

returns the following output,

```
Entropy = 7.999553 bits per byte.

Optimum compression would reduce the size
of this 27587542 byte file by 0 percent.

Chi square distribution for 27587542 samples is 17302.23, and
randomly would exceed this value 0.01 percent of the times.

Arithmetic mean value of data bytes is 126.5112
(127.5 = random).
Monte Carlo value for Pi is 3.159838040 (error 0.58 percent).
Serial correlation coefficient is 0.000521 (totally
uncorrelated = 0.0).
```

which is somewhat verbose but tells us the results of five tests: the entropy of the file, the p-value of the χ^2 test (see Sect. 4.1.1), the arithmetic mean of the bytes of the file, a simulation of the value of π (see Sect. 2.1) and the serial correlation coefficient (see Sect. 4.1.6). We can make the output more compact by adding the -t flag but that output does not contain the p-value for the χ^2 test.

For the TV data the results point towards randomness. The maximum entropy is nearly 8 bits/byte and the χ^2 p-value is close to the range we used previously, $0.05 < p < 0.95$. The estimate of π is 0.6% off and the serial correlation coefficient is quite close to zero.

Can we be more quantitative with the ent results? Perhaps. The five tests can be cast as a vector and then compared to the best possible such vector in a fashion similar to what we did in Sect. 4.2. For example, since we know the best possible entropy, a measure of the information content in the bytes of the file, is 8.0 we can track the deviation from this value and know that the smaller it is the more random the data by this metric. Similarly, the arithmetic mean should approach 127.5 ($255 \div 2$) so the absolute deviation from this value, divided by 127.5, will be a fraction where smaller is better. We can construct a similar deviation for the estimate of π. The serial correlation coefficient can be used as-is, zero is the best possible value. This leaves the χ^2 test results. For this test we can simply define 0 as within the range $0.05 < p < 0.95$ and 1 for anything outside this range.

Putting all of this together defines a five element vector as,

$$(8.0 - e, 0|1, \frac{|m - 127.5|}{127.5}, \frac{|p - \pi|}{\pi}, |c|)$$

where e is the entropy, $0|1$ is from the χ^2 p-value, m is the arithmetic mean, p is the estimate of π, and c is the serial correlation coefficient. We will interpret the norm of this vector as a distance from the results of a perfectly random input. This then gives us a ranking mechanism for comparing pseudorandom generators as we did in Sect. 4.3 as well as the files of random bytes we used previously. For each of the pseudorandom generators from the previous sections we generated 20 output files, each with a different initial seed, and calculated the mean vector across all 20 files. The norm of this vector is then our distance from "perfectly random".

Putting all of this together gives us a ranking of the pseudorandom generators of Sect. 4.3 along with the random data files,

	Score
RANDOM.ORG	0.0000615
xorshift128+	0.0001422
KISS	0.0002028
CMWC	0.0002053
xorshift128	0.0002152
xorshift1024*	0.0500003
Middle Weyl	0.0500005
xorshift32	0.1000002
Mersenne Twister	0.2000002
HotBits	1.0000001
TV	1.0000471
MINSTD (new)	1.0149898
Combined L'Ecuyer	1.0149949
MINSTD (orig)	1.0150025
RANDU	1.0164892

The ranking is satisfying in the sense that the large files of random bytes are at the top. There is a jump in the scores (smaller is better) when the χ^2 test has failed. Interestingly, we see that, according to ent, the HotBits file, 11 MB in size, has failed the χ^2 test. The ordering of the pseudorandom generators is not the same as in the more detailed test suites but as expected RANDU is the worst and the LCG generators are at the bottom, so the trend is believable.

From these results it is clear that the ent program is a useful "quick and dirty" way to gain some intuition on the randomness of a file, especially if no more data can be collected, but care must be taken if the size of the file is too small. As a rule of thumb, if a generator or source of randomness fails the ent tests it will most likely fail the more detailed tests of the previous sections.

4.6 Chapter Summary

In this chapter we explored statistical and empirical tests for randomness and applied them to pseudorandom number generators as well as files of bytes from truly random sources. We started with the classical randomness tests of Knuth and implemented them in C. We then applied these tests to a collection of generators, particularly those of Chap. 2. During this process we learned that it is difficult to declare a file as random or a pseudorandom generator as "good". We then explored two random number test suites in common use, dieharder and TestU01, and applied them to our collection of generators. We also showed how to extend the test suites to accommodate new generators. Lastly, we investigated the ent tool as a means for testing files of bytes for randomness and saw how we could use it to rank the generators resulting in an ordering in line with the more complex and sophisticated test suites.

Exercises

4.1 Section 4.1.2 implements the Kolmogorov-Smirnov test for a small set of samples from a file. Using too large of a sample at one time results in a weak decision. For a single set of n samples, testing for uniformity implies comparing the empirical distribution to $F(x) = x$. If instead, a collection of $m = 1000\ K^{+}_{n=1000}$ (and $K^{-}_{n=1000}$) samples is made from successive $n = 1000$ samples, the KS-test can be applied to that set as well comparing its empirical distribution to $F(x) = 1 - e^{-2x^2}$ to arrive at a single $K^{+}_{m=1000}$ (and $K^{-}_{m=1000}$) value which will be sensitive to both local and global features of the samples. Extend the code in Sect. 4.1.2 to implement this two-tiered test. **

4.2 Section 4.1.3 introduced the serial test for pairs. Extend the code to work with triplets. This means using a three dimensional histogram. Make the number of bins small enough that there will be sufficient counts in each bin. What that means is difficult to quantify, exactly, but each bin should have several dozen values. Apply this to test to the RANDU linear congruential generator of Sect. 2.2. Does it pass? Why or why not?

4.3 Section 4.2 ran the classical randomness tests of Knuth on multiple files from different generators, each time using a different seed. The NIST random excursions test (Sect. 4.1.8) was not included because the classical tests were run on double precision floating-point samples. The generators tested actually output unsigned integer data, either 32-bit or 64-bit (except for the MINSTD generators which should be ignored here). Repeat the failure analysis of Sect. 4.2 for the random excursions test and rank the generators relative to one another. Is this ranking different from the ranking in Sect. 4.2? **

4.4 Locate files of at least 1 million, preferably 10 million or more, bytes and test them with the ent program (Sect. 4.5) using the ranking system developed there. Attempt to use your intuition to select files that are as "random" as possible. How do the selected files compare to the pseudorandom number generators tested?

4.5 Linux systems supply random bytes through /dev/urandom. Write a program to extract 50 files of 20 million double precision floating-point values derived from unsigned 32-bit integers and apply the classical randomness tests of Sect. 4.2. Compare the performance with the generators tested there.

4.6 Using the code in Sect. 4.4, extend TestU01 to test the output of /dev/urandom as was done for the Middle Weyl generator. Run this against the SmallCrush, Crush and BigCrush test suites and comment on the results.

4.7 Write a program to extract a 100 million byte sample from /dev/urandom and rank it with the ent program as in Sect. 4.5.

4.8 This chapter discusses the testing of uniform pseudorandom number generators. In the literature, much has been written about this, but very little has been written about the testing of nonuniform pseudorandom number generators (see Chap. 3). Why?

References

1. Brown, Robert G., Dirk Eddelbuettel, and David Bauer. "Dieharder: A random number test suite." Open Source software library, under development (2013). http://webhome.phy.duke.edu/~rgb/General/dieharder.php.
2. L'Ecuyer, Pierre, and Richard Simard. "TestU01: AC library for empirical testing of random number generators." ACM Transactions on Mathematical Software (TOMS) 33, no. 4 (2007): 22. http://simul.iro.umontreal.ca/testu01/tu01.html.
3. Knuth, D.E. Seminumerical Algorithms. The Art of Computer Programming (2nd ed.), Vol. 2, Addison-Wesley, Reading, MA (1981).
4. Bassham III, Lawrence E., Andrew L. Rukhin, Juan Soto, James R. Nechvatal, Miles E. Smid, Elaine B. Barker, Stefan D. Leigh, et al. "Sp 800-22 rev. 1a. a statistical test suite for random and pseudorandom number generators for cryptographic applications." (2010).
5. https://www.random.org/.
6. https://www.fourmilab.ch/hotbits/.
7. Marsaglia, George. "DIEHARD statistical tests." (CDROM), Florida State University (1995).

Chapter 5
Parallel Random Number Generators

Abstract It is often the case that many separate threads or processes require independent streams of pseudorandom numbers. This chapter examines five methods for generating such streams: a pseudorandom number server process, separate per stream generators, partitioning a single stream into non-overlapping segments, random seeding that relies on the small likelihood of randomly picking overlapping streams, and merging two randomly initialized generator outputs. Implementations for certain generators will be developed and output streams tested.

5.1 Methods for Generating and Testing Streams of Random Numbers

The generators we developed in Chap. 2 all generate a single stream of pseudorandom values from an initial seed or state. Modern computer systems are multithreaded and have many processes running simultaneously, and many of these processes are running their own separate threads. And all of this is to say nothing of the extreme parallelism found in graphics processors. This begs the question: how do we use a single stream of pseudorandom numbers with multiple threads or processes?

Historically, there have been four answers to this question:

1. Use a random number server process
2. Use separate pseudorandom number generator types per stream
3. Partitioning the pseudorandom stream into non-overlapping segments
4. Random seeding of a single generator type per stream

We will examine the last three approaches in the sections below. We will add-in a final section that talks about counter-based generators as these are well-suited to parallel use in two different ways, numbers 1 and 2 above. In addition, we will implement a suggestion by Fog [8] that is similar to random seeding but uses the XOR-ed output of two generators, each initialized with a separate seed.

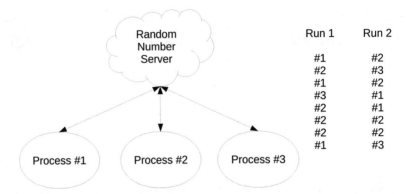

Fig. 5.1 A pseudorandom number server is accessed by three separate processes. Since the order in which the processes request values from the server is not deterministic from Run 1 to Run 2, even if the server process is seeded with the same value between runs, the processes will not be able to repeat prior sequences

The first approach above is quite feasible but is seldom implemented for several reasons. One reason is performance. A server process requires interprocess communication and locking, all of which adds considerable overhead to what is, typically, a very simple set of calculations. For example, the xorshift family of generators involve only a few lines of code. Even setting up interprocess communication involves more code than the generator itself.

An even bigger negative to a random number server is the lack of determinism inherent in the server. Figure 5.1 shows a cartoon of the random number server process. We can assume the server starts with a seed of some kind, like the current epoch system time, and then serves pseudorandom values on demand to every process connecting to it. There is a single stream of pseudorandom numbers because we have a single algorithm using a single seed. However, the *order* in which the separate processes request values is not deterministic as seen in the Fig. 5.1 for Run 1 and Run 2. The ability to seed a generator and produce a repeatable sequence of values is very important in many applications. In this scenario, every run of the processes using the server will result in a different sequence of values served to it.

Parallel generation, conceptualized in this chapter via streams of pseudorandom numbers, leads to another question: how do we test the outputs? We learned in Chap. 4 how to test the output from a single generator. We will test the output of parallel generators in a similar fashion by interleaving the streams, sample by sample . This is exactly the approach taken by the Scalable Parallel Random Number Generators (SPRNG) library [1]. Figure 5.2 shows the process visually.

In the figure, we first generate multiple streams according to one of the options in the following sections. We then interleave the streams to form a single combined output file. It is this file that we then test using any of the pseudorandom number tests of Chap. 4. This approach is effective because any biases introduced by the parallel generation methods will leave a mark in the sequence of values generated by the

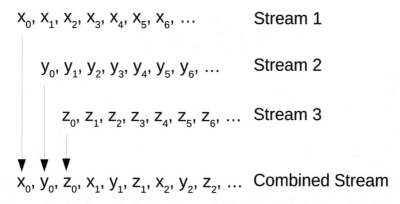

Fig. 5.2 Interleaving the samples from three independent streams. The output file is simply the aggregation of the three streams taken in sets of three. The combined stream is tested with the dieharder or TestU01 framework

set of streams. Even if an individual stream passes all randomness tests it does not immediately imply that the combination of multiple streams will do so. Combining the stream outputs will capture any introduced bias and the tests of Chap. 4 are, by design, intended to reveal any such bias in the output values.

So, then, the parallel generation methods of this chapter will be tested with a simple algorithm,

1. Generate n streams of m random numbers, stored in n files, using any of the methods below.
2. Interleave the n files according to Fig. 5.2 to produce a single output file with nm 32-bit unsigned integers.
3. Pass this single output file to dieharder and evaluate the output by looking at the number of passing, weak, and failing tests.

The program to interleave n streams of m samples is straightforward in Python,

```
 1  import os
 2  import sys
 3
 4  def main():
 5      """Interleave a set of streams"""
 6
 7      oname = sys.argv[1]
 8      fnames = sys.argv[2:]
 9      g = open(oname, "w")
10      files = []
11      for fname in fnames:
12          files.append(open(fname))
13      b = os.stat(fnames[0]).st_size
14      nsamp = b/4
```

```
15        for i in xrange(nsamp):
16            for k in range(len(files)):
17                c = files[k].read(4)
18                g.write(c)
19        g.close()
20        for f in files:
21            f.close()
22
23  main()
    (interleave_streams.py)
```

where the first argument is the name of the combined output file and all other arguments are the names of the individual stream output files, each assumed for simplicity to be of the same length. If the stream files are not all the same length, place the shortest stream file first. We then open the files (lines 9–12), determine how many 32-bit samples we will read from each file (lines 13–14) and then read that many samples, across each stream file, outputting to the combined file each time (lines 15–18).

Let's be more specific about how we will simulate the streams. We need not complicate matters by creating multithreaded or multiprocess applications. Instead, we simulate streams by creating multiple instances of Python classes each of which encapsulates a pseudorandom generator from Chap. 4. For example, we encapsulate the MINSTD generator with,

```
1   class MINSTD:
2       def next(self):
3           self.x = (self.A*self.x) % self.M
4           return self.x
5       def next_fp(self):
6           return float(self.next()) * 4.656612875245797e-10
7       def jump(self, p):
8           def mod_pow(base, exponent, modulus):
9               if modulus == 1:
10                  return 0
11              c = 1
12              for i in xrange(1, exponent+1):
13                  c = (c * base) % modulus
14              return c
15          ap = mod_pow(self.A, p, self.M)
16          self.x = (ap * self.x) % self.M
17      def init(self, seed=12345):
18          self.x = seed
19      def __init__(self, seed=12345, A=48271, M=2147483647):
20          self.A = A
21          self.M = M
22          self.init(seed)
    (minstd.py)
```

where we define four methods, which will be common to most of the Python classes: init, jump, next_fp and next. We use the init method to re-initialize the object with a different seed. The constructor has the option of accepting a seed as well as new A and M constants, if we wish. The default constants define the revised MINSTD generator of Park and Miller [2]. Changing A to 16807 will instantiate the original version of the generator.

The next method implements the LCG calculation. We do not use Schrage decomposition here because Python readily supports large integer calculations. The next_fp method takes the output of next and returns it as a floating point value in [0, 1). We will ignore the jump method for the moment. It is used in Sect. 5.3 and we will describe it there.

The MINSTD class is simple to use,

```
>>> from minstd import *
>>> g = MINSTD()
>>> [g.next() for i in range(4)]
[595905495, 1558181227L, 1498755989L, 2021244883L]
>>> [g.next_fp() for i in range(4)]
[0.41314081401244773, 0.7202331948653949, 0.3765493474791522,
    0.4135521661553309]
>>> g.init()
>>> [g.next() for i in range(4)]
[595905495, 1558181227L, 1498755989L, 2021244883L]
>>>
```

The other Python classes will be introduced when they are used in the sections below.

5.2 Per Stream Generators

One approach to parallel generators is to use a different generator for each stream. If the number of streams is small and the number of available (good) generators is at least as large, then this approach might work well. One approach is to use a common base generator for each stream but to vary the parameters of the generator so that each stream will create a different sequence of random values. This approach should not be confused with simply using the same generator for each stream with a different starting seed. We will consider that approach in Sect. 5.4.

For this example, we will use the xorshift32 generator [3] as it is parameterized by a triplet of constants, (a, b, c). In the original paper Marsaglia gives 81 such triplets. We will use these to create up to 81 independent streams like so,

```
 1 import random
 2 xorshift32_PARAMS = [
 3     [ 1,  3,10],[ 1,  5,16],[ 1,  5,19],[ 1,  9,29],[ 1,11,  6],
       [ 1,11,16],[ 1,19,  3],[ 1,21,20],[ 1,27,27],
 4     [ 2,  5,15],[ 2,  5,21],[ 2,  7,  7],[ 2,  7,  9],[ 2,  7,25],
       [ 2,  9,15],[ 2,15,17],[ 2,15,25],[ 2,21,  9],
 5     [ 3,  1,14],[ 3,  3,26],[ 3,  3,28],[ 3,  3,29],[ 3,  5,20],
       [ 3,  5,22],[ 3,  5,25],[ 3,  7,29],[ 3,13,  7],
 6     [ 3,23,25],[ 3,25,24],[ 3,27,11],[ 4,  3,17],[ 4,  3,27],
       [ 4,  5,15],[ 5,  3,21],[ 5,  7,22],[ 5,  9,7 ],
 7     [ 5,  9,28],[ 5,  9,31],[ 5,13,  6],[ 5,15,17],[ 5,17,13],
       [ 5,21,12],[ 5,27,  8],[ 5,27,21],[ 5,27,25],
 8     [ 5,27,28],[ 6,  1,11],[ 6,  3,17],[ 6,17,  9],[ 6,21,  7],
       [ 6,21,13],[ 7,  1,  9],[ 7,  1,18],[ 7,  1,25],
 9     [ 7,13,25],[ 7,17,21],[ 7,25,12],[ 7,25,20],[ 8,  7,23],
       [ 8,9,23 ],[ 9,  5,1 ],[ 9,  5,25],[ 9,11,19],
10     [ 9,21,16],[10,  9,21],[10,  9,25],[11,  7,12],[11,  7,16],
       [11,17,13],[11,21,13],[12,  9,23],[13,  3,17],
11     [13,  3,27],[13,  5,19],[13,17,15],[14,  1,15],[14,13,15],
       [15,  1,29],[17,15,20],[17,15,23],[17,15,26]]
12
13 class xorshift32:
14     def next(self):
15         MASK = 0xffffffff
16         self.x ^= (self.x << self.A) & MASK
17         self.x ^= (self.x >> self.B) & MASK
18         self.x ^= (self.x << self.C) & MASK
19         return self.x
20     def next_fp(self):
21         return float(self.next()) * 2.3283064365386963e-10
22     def init(self, seed=12345):
23         self.x = seed
24     def __init__(self, seed=12345, stream=0):
25         global xorshift32_PARAMS
26         self.stream = stream
27         self.A, self.B, self.C = xorshift32_PARAMS
                 [stream % len(xorshift32_PARAMS)]
28         self.init(seed)
29
30 random.shuffle(xorshift32_PARAMS)
   (xorshift32.py)
```

When the module is loaded,

```
>>> from xorshift32 import *
```

line 30 will execute to shuffle the order of the parameter list, xorshift32_PARAMS. This will make each run use a different sequence of parameters per stream. Naturally, one might wish to use the same sequence in which case line 30 can be removed or replaced with some other initialization function that defines a specific order for this set of parameters. In a sense, a hyperseed for the generators.

The xorshift32 class is a straightforward implementation of the C version with the addition of MASK to keep operations in the proper 32-bit range. The constructor accepts two arguments, a seed, and a stream number. The seed can be reset by calling init but the stream, which set of parameters from xorshift32_PARAMS this instance is using, cannot be changed after the object is instantiated. It is up to the user to ensure that stream is used properly to keep the output of the objects separate.

Using the class is straightforward,

```
>>> from xorshift32 import *
>>> x = xorshift32(stream=0)
>>> y = xorshift32(stream=1)
>>> z = xorshift32(stream=2)
>>> [x.next() for i in range(4)]
[184851L, 3983261L, 40832471L, 859415511L]
>>> [y.next() for i in range(4)]
[25951528L, 226639114L, 3453807590L, 3872092074L]
>>> [z.next() for i in range(4)]
[410042495L, 2247877535L, 4260414942L, 35950270L]
```

where x, y, and z are separate streams each using a different set of (a, b, c) constants. Note that all three streams are using the same default starting seed, the air shield combination value of 12345.

Let's write a small driver program to generate streams of xorshift32 samples that we can test by combining them with the interleave program in Sect. 5.1,

```
1  import os
2  import sys
3  import struct
4  import random
5  from xorshift32 import *
6
7  def main():
8      streams = int(sys.argv[1])
9      samples = int(sys.argv[2])
10     seedtype= int(sys.argv[3])
11     base    = sys.argv[4]
12     g = []
13     for i in range(streams):
14         if (seedtype==1):
15             s = xorshift32
                   (seed=int((2**32-1)*random.random()),
                    stream=i)
16         else:
17             s = xorshift32(stream=i)
18         g.append(s)
```

```
19       for i in range(streams):
20           with open(base+("_%02d.dat" % i),"w") as f:
21               for j in range(samples):
22                   n = g[i].next()
23                   f.write(struct.pack("L",n))
24
25 main()
    (x32_streams.py)
```

We create as many instances of xorshift32 as we need using the default seed (line 17) or a randomly selected seed (line 15). Recall that Python's native random number generator is the Mersenne Twister. We ran the above to produce five output streams, each with 100 million samples, first with the default xorshift32 seed and then with a randomly selected seed. We then subjected each of the merged streams (using the interleave script above) to the dieharder test suite. The results are, where "Score" is $2p + w - 2f$ for p passing, w weak, and f failing tests,

Seed type	Passed	Weak	Failed	Score
Fixed	101	4	9	188
	105	6	3	210
	107	5	2	215
	108	4	2	216
	99	10	5	198
Random	106	5	3	211
	102	6	6	198
	103	5	6	199
	100	5	9	187
	107	4	3	212

indicating that for this experiment there was no difference in performance between using the same seed for each stream or using separate seeds for each stream. This is not too surprising given that the different set of (a, b, c) parameters in the different xorshift32 streams would be expected to lead to very different output sequences.

The xorshift32 generator is an integral component of the KISS32 generator. Therefore, we can extend KISS to use the stream version of xorshift32 and thereby make this long period generator accessible to streaming. If we do this, KISS32 becomes, in Python,

```
1 from xorshift32 import *
2 class KISS:
3     def lcg32(self):
4         self.cng = ((69069*self.cng)&0xffffffff) + 13579
5         self.cng &= 0xffffffff
6         return self.cng
```

```
 7      def mwc32(self):
 8          MASK = 0xffffffff
 9          self.j = ((self.j+1)&4194303)&MASK
10          x = self.Q[self.j]
11          t = (((x<<28)&MASK)+self.carry)&MASK
12          self.carry = ((x>>4)&MASK)-(t<x)
13          self.Q[self.j] = t-x
14          return t-x
15      def next(self):
16          n = self.mwc32() + self.x32.next() + self.lcg32()
17          n &= 0xffffffff
18          return n
19      def next_fp(self):
20          return float(self.next()) * 2.3283064359965952e-10
21      def __init__(self, seed=12345, stream=0):
22          self.cng = 123456789
23          self.x32 = xorshift32(seed, stream=stream)
24          self.carry = 0
25          self.Q = [0]*4194304
26          self.j = 4194303
27          for i in xrange(4194304):
28              self.Q[i] = self.lcg32() + self.x32.next()
```
(kiss.py)

with the constructor initializing the generator for the given stream which is applied to the xorshift32 portion. Note that the implementation is for teaching purposes only, it is too slow to initialize for serious use, which would likely switch to the NumPy library from pure Python.

If we use this generator to create five interleaved output files, each from five separate streams using five separate parameterizations of xorshift32, we can test them with dieharder. The mean test score is 223.4 with only a single test failure clearly indicating that the quality of the KISS generator is not affected by using a stream version of xorshift32.

5.3 Skipping

If we split the output of a single generator into n non-overlapping subsequences, each m samples long, we would then be assured of having truly independent streams of random numbers as long as our processes consumed less than m samples. We could do this by repeated calls to the generator, of course. We seed the generator, call it m times storing the output somewhere, and call this output "stream 1". Then, without reseeding, we call the generator another m times, store that output, and call it "stream 2", and so on until we have output for n streams. This is highly inefficient because it requires nm calls to the generator to say nothing of the overhead of storing all those samples and creating code to quickly use them when needed by the proper stream. What would be nice would be to advance the generator some number of samples, say m, *without* generating all the samples in-between. In that case, we only need to advance the generator im samples where m is the number of samples per stream and i is the current stream number starting with zero. Doing just this is what this section is about.

Not all generators can be advanced quickly. For the KISS generator, for example, jumping from sample x_n to x_{n+m} for some m is not simple at all, if even possible. However, some generators can be advanced quickly, either by a fixed but large number of steps or by an arbitrary number of steps. Here we will look at three such generators: xorshift128+, xorshift1024* and MINSTD. The first two are improved versions of the original xorshift generator (see Sect. 2.4) for which jump functions exist to advance the output by a large, fixed value. The MINSTD generator, as a linear congruential generator, can be advanced by an arbitrary amount. With these generators we can output streams of non-overlapping samples, interleave them as in the previous section, and test them with dieharder. But first, we need implementations of the generators that are amenable to streamed output along with an understanding of how the jump ahead works.

Let's start with MINSTD. The basic operation of a linear congruential generator is,

$$x_1 = ax_0 \bmod m$$

where for simplicity we are using $c = 0$ (a multiplicative linear congruential generator). Expanding this one step further gives,

$$x_2 = ax_1 \bmod m = a(ax_0) \bmod m = a^2 x_0 \bmod m$$

which implies that if we are at x_n and we want to advance m positions we need only do,

$$x_{n+m} = a^m x_n \bmod m$$

which seems straightforward enough. Note that the a^m we need is really $a^m \bmod m$. This is good as without the modulo operation a^m will quickly become very large indeed for the m values we would like to use. For example, $m = 10^6$ would be a small stream. How do we calculate $a^m \bmod m$? The key is to see that we do the modulo after every $a \times a$ operation,

$$a^3 \bmod m = a(a^2 \bmod m) \bmod m$$

giving us an algorithm for finding the proper value.

In Sect. 5.1 we gave an implementation of MINSTD in Python with a `jump` method which we intentionally neglected to explain at the time. We will explain it now. The method is,

```
 7  def jump(self, p):
 8      def mod_pow(base, exponent, modulus):
 9          if modulus == 1:
10              return 0
11          c = 1
12          for i in xrange(1, exponent+1):
13              c = (c * base) % modulus
14          return c
15      ap = mod_pow(self.A, p, self.M)
16      self.x = (ap * self.x) % self.M
```

where we are keeping the line numbers for easy comparison with the listing above. The method includes a nested function, mod_pow, which implements the modular exponent we just derived. The actual jump which advances the state of the generator is in lines 15 and 16. We first calculate a^p mod m to advance by p samples and then do the actual advance in line 16 by multiplying the current state by a^p.

A few statements will test the jump method to show that it is indeed advancing the generator,

```
>>> x = MINSTD()
>>> for i in xrange(10000000):
...     v=x.next()
...
>>> [x.next() for i in range(4)]
[308157786L, 1612748884L, 571692167L, 987729307L]

>>> x = MINSTD()
>>> x.jump(10000000)
>>> [x.next() for i in range(4)]
[308157786L, 1612748884L, 571692167L, 987729307L]
```

where in the first block of code we instantiate a generator and then explicitly call next ten million times. We then show the next four calls to next. In the second block we instantiate the generator, with the same seed as before, and call jump to advance ten million times. The next four calls to next show that we are indeed in the same place in the sequence.

Let's define the xorshift128+ and xorshift1024* generators. In Python these are,

```
 1  import struct
 2  class xorshift128plus:
 3      def next(self):
 4          MASK = 0xffffffffffffffff
 5          s1 = self.x[0]
 6          s0 = self.x[1]
 7          result = (s0+s1) }& MASK
 8          self.x[0] = s0 }& MASK
 9          s1 ^= s1 << 23
10          self.x[1] = (s1 ^ s0 ^ (s1 >> 18) ^ (s0 >> 5)) }&
                MASK
11          return result
```

```
12      def next_fp(self):
13          n = (self.next() >> 12) | 0x3FF0000000000000
14          d = struct.unpack("d", struct.pack("Q",n))[0] - 1.0
15          return d
16      def jump(self):
17          JUMP = [ 0x8a5cd789635d2dff, 0x121fd2155c472f96]
18          s0 = s1 = 0
19          for i in range(2):
20              for b in range(64):
21                  if (JUMP[i] }& (1<<b)):
22                      s0 ^= self.x[0]
23                      s1 ^= self.x[1]
24                  self.next()
25          self.x = [s0,s1]
26      def init(self, seed=12345):
27          self.x = [seed, (seed >> 4) ^ (seed+seed)]
28      def __init__(self, seed=12345):
29          self.init(seed)
```
(xorshift128plus.py)

and,

```
1  import struct
2  class xorshift1024star:
3      def next(self):
4          MASK = 0xffffffffffffffff
5          s0 = self.x[self.p]
6          self.p = (self.p+1) }& 15
7          s1 = self.x[self.p]
8          s1 ^= (s1 <<31) }& MASK
9          self.x[self.p] =   \
10             (s1 ^ s0 ^ (s1 >> 11) ^ (s0 >> 30)) }& MASK
11         res = self.x[self.p]
12         res = (res * 1181783497276652981) }& MASK
13         return res
14     def next_fp(self):
15         n = (self.next() >> 12) | 0x3FF0000000000000
16         d = struct.unpack("d", struct.pack("Q",n))[0] - 1.0
17         return d
18     def jump(self):
19         JUMP = [
20             0x84242f96eca9c41d,0xa3c65b8776f96855,
                   0x5b34a39f070b5837,
21             0x4489affce4f31a1e,0x2ffeeb0a48316f40,
                   0xdc2d9891fe68c022,
22             0x3659132bb12fea70,0xaac17d8efa43cab8,
                   0xc4cb815590989b13,
23             0x5ee975283d71c93b,0x691548c86c1bd540,
                   0x7910c41d10a1e6a5,
```

```
24                    0x0b5fc64563b3e2a8,0x047f7684e9fc949d,
                          0xb99181f2d8f685ca,
25                    0x284600e3f30e38c3]
26        t = [0]*16
27        for i in range(2):
28            for b in range(64):
29                if (JUMP[i] }& (1<<b)):
30                    for j in range(16):
31                        t[j] ^= self.x[(j+self.p) }& 15]
32                    self.next()
33            for j in range(16):
34                self.x[(j+self.p) }& 15] = t[j]
35    def init(self, seed=12345):
36        MASK = 0xffffffffffffffff
37        self.x = [0]*16
38        self.p = 0
39        self.x[0] = seed }& MASK
40        for i in range(1,16):
41            self.x[i] = (self.x[i-1]*seed + seed) }& MASK
42    def __init__(self, seed=12345):
43        self.init(seed)
```
(xorshift1024star.py)

Both of these classes define the expected `next`, `next_fp`, and `init` methods. Since these generators produce unsigned 64-bit output we use the same trick we did in Chap. 2 to define the floating-point value returned by `next_fp`: explicitly defining the IEEE 754 double precision bits and then using the Python `struct` library module to convert directly to a float. This makes full use of all 53 bits in the significand of the IEEE double precision format. See Sect. 2.4. The `next` and `init` methods are direct Python versions of the C code given in Sect. 2.4.

The `jump` methods are new. For xorshift128+ `jump` advances the state by 2^{64} calls meaning that for any given seed one can in theory create 2^{64} streams each with 2^{64} samples. For xorshift1024* `jump` advances by 2^{512} samples enabling up to 2^{512} streams of 2^{512} samples each. Either of these options are more than sufficient for almost any use of the generators.

The `jump` methods are really calculating vT^j where v is the current state (a 64 bit vector), T is the transition matrix (see Sect. 2.4) and j is some integer constant, the number of calls to jump. Remember that the operations of an xorshift generator are equivalent to vT so that for a starting seed value, v, the first output value is vT, the second is vT^2, the third is vT^3, and so on. The `jump` method calculates for a fixed j, either $j = 2^{64}$ or $j = 2^{512}$, a representation of T^j through the characteristic polynomial so that we have,

$$vT^j = \sum_{i=0}^{n} \alpha_i vT^i$$

with n the number of state vectors and the α values, which are bit values, 0 or 1, found ahead of time for the selected j value. This explains the "magic" constants in the `jump` methods. Recall that the xorshift operations happen in F_2 so that the XOR operation is addition.

A driver program,

```
1   import sys
2   import struct
3   import random
4   from minstd import *
5   from xorshift128plus import *
6   from xorshift1024star import *
7   def main():
8       gen = int(sys.argv[1])
9       streams = int(sys.argv[2])
10      samples = int(sys.argv[3])
11      base = sys.argv[4]
12      g = []
13      if (gen==0):
14          seed = int((2**31-1)*random.random())
15      else:
16          seed = int((2**64-1)*random.random())
17      for i in range(streams):
18          if (gen == 0):
19              s = MINSTD()
20              s.init(seed)
21              s.jump(int(1.1*samples*i))
22              g.append(s)
23          elif (gen == 1):
24              s = xorshift128plus()
25              s.init(seed)
26              for j in range(i):
27                  s.jump()
28              g.append(s)
29          elif (gen == 2):
30              s = xorshift1024star()
31              s.init(seed)
32              for j in range(i):
33                  s.jump()
34              g.append(s)
35      for i in range(streams):
36          with open(base+("_%03d.dat" % i), "w") as f:
37              if (gen == 0):
38                  for j in range(samples):
39                      n = int((2**32)*g[i].next_fp())
40                      f.write(struct.pack("L",n))
```

```
41              elif (gen == 1):
42                  for j in range(samples/2):
43                      n = g[i].next()
44                      f.write(struct.pack("Q",n))
45              elif (gen == 2):
46                  for j in range(samples/2):
47                      n = g[i].next()
48                      f.write(struct.pack("Q",n))
49 main()
   (skipping.py)
```

will generate output stream files for MINSTD, xorshift128+, or xorshift1024* which, when merged, can be tested. Note that when actually generating samples for MINSTD we use the floating-point output and scale it by 2^{32} to get an unsigned 32-bit output value. For the xorshift generators we get a 64-bit output that we store as two 32-bit outputs explaining why we only generate half as many samples as requested. Also note that we use *the same seed* (lines 13–16) for each stream and partition the streams by calling the respective jump method. For MINSTD we jump by 110% of the number of samples we need to ensure no overlap between the streams.

For this experiment we generated 10 separate streams of 50 million samples for each generator and then interleaved them to create a single test file for dieharder. The results were,

	Passed	Weak	Failed	Score
MINSTD	94	8	12	172
xorshift128+	110	4	0	224
xorshift1024*	112	2	0	226

The xorshift generator results were essentially the same in this case as those of Sect. 4.3. This is encouraging. However, the MINSTD score of 172 is quite a bit lower than the single stream score of 200 in Sect. 4.3. To see if this is a statistical fluke or a true result, we repeated the experiment ten times and compared the mean score with that found by using the dieharder built-in version of MINSTD, also run ten times. The results are (mean ± SE) 200.000 ± 0.789 for dieharder's MINSTD and 167.273 ± 0.982 for the multistream version. A *t*-test gives a *p*-value of less than 0.000001 likelihood that this difference is not real. So, clearly, skipping with MINSTD is not working as nicely as skipping with the xorshift generators. Yet another strike against linear congruential generators.

5.4 Random Seeding

Our next approach to parallel pseudorandom number generation is to play the odds and simply initialize a generator with seeds generated by a separate generator. This approach uses the same underlying generator each time but is not skipping ahead by any amount. Instead, we are simply hoping that we will land in sufficiently different parts of the space of sequences output by the generator to be immune to overlap or other spurious correlations between the streams. This method is simple and also attractive because it will work for any pair of generators, not just those that can be parameterized or skipped. Random seeding was used in [4] to successfully apply two-dimensional Gaussian curve fitting to hundreds of thousands of small images using a hybrid Tausworthe [5] generator for seeding and MINSTD running per thread on the GPU. The selection of generators in this case was not optimal but still successful because of the relatively few values needed to fit each Gaussian. We will see better pairings below.

The probability of s streams of length l from a single generator with period ρ and random seeds *not* overlapping anywhere is approximately,

$$p = (1 - sl/\rho)^{s-1}$$

as given in [6]. This formula implies that the single generator should have as large a period as possible since $sl/\rho \to 0$ as $\rho \to \infty$.

We can easily compare pairs of generators with a simple program that creates streams using a source generator for seeds and a destination generator that uses one of those seeds for each stream. Testing follows the approach of previous sections: interleave the streams and pass the output to dieharder to test as if it were a single stream.

In Chap. 2 we implemented many generators in C. We can simulate the streams by generating a single stream with each destination generator changing the seed to another draw from the source generator after each set of n samples has been generated. Building on the existing C implementations of Chap. 2 we only need a straightforward driver program,

```
1  int main(int argc, char *argv[]) {
2      int i,j,sgen,dgen,nstream,nsamp,sseed;
3      FILE *f;
4      char fname[256];
5      uint32_t *seeds, u32;
6
7      sgen = atoi(argv[1]);
8      dgen = atoi(argv[2]);
9      nstream = atoi(argv[3]);
10     nsamp = atoi(argv[4]);
11     sseed = atoi(argv[5]);
12     switch (sgen) {
13         case 0:  CMWC_init(sseed);       break;
14         case 1:  init_genrand(sseed);    break;
15         case 2:  kiss32_init(sseed);     break;
16         case 3:  mwseed |= sseed;        break;
17         default: s2 = (int32_t)sseed;    break;
18     }
19     seeds = (uint32_t *)malloc(nstream*sizeof(uint32_t));
```

```
20        for(i=0; i<nstream; i++) {
21            switch (sgen) {
22                case 0:   seeds[i]=CMWC();              break;
23                case 1:   seeds[i]=genrand_int32();     break;
24                case 2:   seeds[i]=kiss32();            break;
25                case 3:   seeds[i]=middle_weyl();       break;
26                default:  seeds[i]=uint32_combined_rng2();break;
27            }
28        }
29        for(i=0;  i < nstream;  i++) {
30            sprintf(fname,"stream_%03d.dat",i);
31            f = fopen(fname,"w");
32            switch (dgen) {
33                case 0:   CMWC_init(seeds[i]);         break;
34                case 1:   init_genrand(seeds[i]);      break;
35                case 2:   kiss32_init(seeds[i]);       break;
36                case 3:   mwseed |= seeds[i];          break;
37                default:  s2 = (int32_t)seeds[i];      break;
38            }
39            for(j=0; j<nsamp; j++) {
40                switch (dgen) {
41                    case 0:   u32 = CMWC();             break;
42                    case 1:   u32 = genrand_int32();    break;
43                    case 2:   u32 = kiss32();           break;
44                    case 3:   u32 = middle_weyl();      break;
45                    default:  u32 = uint32_combined_rng2();
                          break;
46                }
47                fwrite((void*)}&u32, sizeof(uint32_t), 1, f);
48            }
49            fclose(f);
50        }
51        return 0;
52    }
```
(random_seeding.c)

This driver program supports five different generators: CMWC, Mersenne Twister, KISS32, Middle Weyl, and L'Ecuyer's combined LCG. We can use any combination as source and/or destination generator. We seed the source generator and then generate all the necessary destination generator seeds, one for each output stream, in lines 12–28. Lines 29–50 seed the destination generator with one of these seeds and then generates nsamp samples as unsigned 32-bit integers for output to disk. This is repeated for each stream. The interleave Python script of Sect. 5.1 will combine these for dieharder.

The combined LCG generator produces output in the range $[0, 2^{31} - 1)$ which does not cover the full 32-bit range. To get around this, we first change the output to a C double by dividing by $2^{31} - 1 = 2147483647$ and then multiply this value by $2^{32} = 4294967296$,

Table 5.1 Scores calculated from dieharder test results

	CMWC	Mersenne Twister	KISS	Middle Weyl	Combined LCG
CMWC	223	221	220	215	175
Mersenne Twister	223	223	225	224	176
KISS	225	223	225	221	177
Middle Weyl	223	222	221	221	178
Combined LCG	218	224	224	220	175

The score is $2p + w - 2f$ where p is the number of passing, w is the number of weak, and f is the number of failing tests. The row gives the source generator for the seeds and the columns are the destination generators for the streams. Ten streams of 50 million unsigned 32-bit integers were interleaved for each pairing and run through dieharder

```
uint32_t uint32_combined_rng2() {
    return (uint32_t)
        (4294967296.0*((double)combined_rng2()/2147483647.0));
}
```

A simple script generates ten streams of 50 million samples for each possible source generator (seeds) and destination generator (output) giving us 25 merged output streams of 500 million samples each. Applying all dieharder tests and our scoring function ($2p+w-2f$) gives Table 5.1 where each row is the seed generator, itself always seeded with 12345, and each column is the destination generator. A perfect score is 228 as there are 114 tests in the dieharder suite.

From the selection of generators used in Table 5.1 it is clear, and expected, that the combined LCG generator is not a good choice for a destination generator. This is in agreement with the intuition we had above that destination generators with longer periods might be better choices for output. The KISS column gives further support to this idea. Tracing the diagonal, all the tests where the seed and destination generators were the same again supports the idea of using KISS for both source and destination though the differences in the scores is quite small. In fact, if the combined LCG is ignored almost any pairing of the remaining four generators leads to good results with the (potentially) curious exception of CMWC seeding Middle Weyl which gave a relatively low score of 215.

Running the CMWC source to Middle Weyl destination tests ten times over gives a mean score of 221.20 with a standard error of 0.92 indicating that the single 215 in Table 5.1 is not representative but on the low end of what one would expect in practice.

Are more subtle interactions buried in Table 5.1? Perhaps, but if they are there they are indeed subtle and might only show themselves by using dieharder's ability to "ramp up" the testing to push generators to their limits. Also, with sufficiently large output files, one might be able to effectively use TestU01's Crush or even Big Crush suites to add a different perspective to the results.

5.5 Fog Method

In [8] several methods are given for producing parallel streams of random samples. Most mirror the methods we have implemented in the sections above but the seventh is one we will implement and test in this section.

This method uses two separate generators which may be of the same type but must be separate instances with separate state. Each stream uses a randomly initialized pair of these generators to produce as many random samples as needed where each output sample is the XOR of the output of the two generators. In [8] it is claimed that these outputs when XOR-ed are robust to initialization even with a quite poor seed generator.

Algorithmically, the steps to produce a stream of output are,

1. Select a seed s_1 for generator G_1.
2. Select a seed s_2 for generator G_2.
3. Each sample requested is G_1 XOR G_2.

The seeds themselves may be selected with any other generator itself initialized with a specific seed if the entire set of stream outputs is to be deterministic.

Let's make this concrete by simulating streams in C for any pair of generators from CMWC, KISS, Mersenne Twister, and Middle Weyl. The seeds will come from either MINSTD or RANDU to test the claim that the quality of the seed generator is not crucial. As the generators need to have different state we will implement each twice and simply add a 0 or 1 to the function names. Then, a stream is a specified number of samples for a specified pair of generators initialized with successive calls to the seed generator.

In code this becomes,

```
 1  void main(int argc, char *argv[]) {
 2      int i,j,zgen,sgen,dgen,nstream,nsamp;
 3      FILE *f;
 4      char fname[256];
 5      uint32_t sseed,u;
 6      sgen = atoi(argv[1]);     dgen = atoi(argv[2]);
 7      nstream = atoi(argv[3]); nsamp = atoi(argv[4]);
 8      zgen  = atoi(argv[5]);    sseed = atoi(argv[6]);
 9      if (zgen==0) seed0 = sseed;
10      else srandu = sseed;
11      for(i=0; i < nstream; i++) {
12          sprintf(fname,"stream_%03d.dat",i);
13          f = fopen(fname,"w");
14          sseed = (zgen) ? randu() : minstd0();
15          switch (dgen) {
16              case 0:  CMWC_init0(sseed);     break;
17              case 1:  init_genrand0(sseed);  break;
18              case 2:  kiss32_init0(sseed);   break;
19              default: mwseed0 |= sseed;      break;
20          }
```

```
21        sseed = (zgen) ? randu() : minstd0();
22        switch (sgen) {
23            case 0:  CMWC_init1(sseed);      break;
24            case 1:  init_genrand1(sseed);   break;
25            case 2:  kiss32_init1(sseed);    break;
26            default: mwseed1 |= sseed;       break;
27        }
28        for(j=0; j<nsamp; j++) {
29            switch (dgen) {
30                case 0:  u = CMWC0();             break;
31                case 1:  u = genrand_int320();   break;
32                case 2:  u = kiss320();          break;
33                default: u = middle_weyl0();     break;
34            }
35            switch (sgen) {
36                case 0:  u ^= CMWC1();             break;
37                case 1:  u ^= genrand_int321();   break;
38                case 2:  u ^= kiss321();          break;
39                default: u ^= middle_weyl1();     break;
40            }
41            fwrite((void*)&u, sizeof(uint32_t), 1, f);
42        }
43        fclose(f);
44    }
45 }
```
(*xor_streams.c*)

where lines 1–8 parse the command line options. Lines 9 and 10 set the seed generator's own seed value from the command line, either MINSTD or RANDU. The `for` loop in line 11 generates each of the streams, here simulated sequentially but in practice using per stream instances of the two selected generators. Note that once seeded, the seed generator is simply called as needed meaning each stream is initialized with two successive samples from the seed generator.

For each stream, lines 12 and 13 set up the output sample file. Lines 14 through 20 seed the first selected generator, note the o suffix to the function name in this example. Lines 21 through 27 do the same for the second generator.

The samples are generated in lines 28 through 42. For each selected generator a sample is drawn (lines 29–34 and 35–40) where the second sample is immediately XOR-ed with the first. Line 41 outputs the combined sample to disk.

If we run this code for each possible pair of generators, aware that the combination of CMWC and MT is the same in practice as MT and CMWC, and for each of the two possible seed generators, we will have streams which, when interleaved as all the streams of this chapter are, can be tested and scored with dieharder. Ultimately, then, we are left with Table 5.2.

From the table it is evident that the Fog method is quite good regardless of the generator pairing or the seed generator used. There is no statistically significant

Table 5.2 Dieharder scores for the Fog method applied to pairs of generators with seeds coming from MINSTD or RANDU

	CMWC	Mersenne Twister	KISS	Middle Weyl
With MINSTD				
CMWC	226	214	223	224
Mersenne Twister		223	224	224
KISS			226	220
Middle Weyl				222
With RANDU				
CMWC	221	226	223	225
Mersenne Twister		219	227	225
KISS			220	224
Middle Weyl				225

Regardless of the pairing or seed generator the results are all quite good (maximum score 228)

difference between using MINSTD or RANDU for the seed generator proving the claim in [8] that this method is robust to the seed source. Additionally, there is no statistically significant difference between Tables 5.2 and 5.1 (ignoring the combined LCG generator) showing the random seeding test results.

5.6 Counter-Based Generators in Parallel

In Chap. 2 we introduced two counter-based generators, Threefry and Philox. These generators are very flexible when used for parallel purposes which justifies their use and makes up for their relatively slow speeds. By altering the key associated with the generator we can implement multiple streams via parameterized versions copying the approach of Sect. 5.2. Alternatively, by fixing the key and simply changing the range of indices into the generators we can easily partition the sequence and implement the approach of Sect. 5.3 without the need for jump functions.

Let's vary the key which is the same as varying the seed as each key leads to a very different sequence of output values that is highly uncorrelated. It is understood that because this is close to simply varying the seed of other generators calling this a parameter is a bit of semantics. Nevertheless, we will proceed and see where it leads us.

The source code for the Philox multistream generator is straightforward if we use the code in Sect. 2.6 which configures the Philox generator to follow the form of other generators with a seed (a key) and successive calls to a single function moving sequentially through the sequence generated by the key. This gives us,

```
 1  void main(int argc, char *argv[]) {
 2      FILE *f;
 3      char fname[256];
 4      uint32_t i,j, nsamp, nstreams, u;
 5
 6      if (argc == 1) {
 7          printf("\nphilox_streams_param <nstreams> <nsamp>
                <seed>\n\n");
 8          printf("  <nstreams> - number of streams
                to generate\n");
 9          printf("  <nsamp>    - number of uint32 samples
                per stream\n");
10          printf("  <seed>     - key generator seed
                value\n\n");
11          return;
12      }
13
14      nstreams = (uint32_t)atol(argv[1]);
15      nsamp = (uint32_t)atol(argv[2]);
16      seed = (uint32_t)atol(argv[3]);
17
18      for(i=0; i<nstreams; i++) {
19          sprintf(fname,"stream_%03d.dat",i);
20          f = fopen(fname,"w");
21          philox32_init(xorshift32());
22          for(j=0; j<nsamp; j++) {
23              u = philox32();
24              fwrite((void*)&u, sizeof(uint32_t), 1, f);
25          }
26          fclose(f);
27      }
28  }
```
(philox_streams_param.c)

where lines 1–16 simply read command line arguments for the number of streams
(nstreams), number of samples per stream (nsamp), and the seed for the xorshift32
generator that sets the keys for each stream (seed). This should not be confused
with the keys themselves. We simply add it as an option so that each sequence of
stream values can be repeated, if desired.

The loop of lines 18–27 generates a single stream of samples with a unique key.
Line 21 sets the key. The loop of lines 22–25 outputs the samples for the current
stream. Each call to philox32 returns the next 32-bit value from the sequence
specified by the key. Line 24 writes this sample to disk until the stream is complete.
The code for Threefry is identical but writes a single 64-bit value to disk (see
threefry_streams_param.c).

Table 5.3 Counter-based generators in parallel by varying the key

Generator	Passed	Weak	Failed	Score
(a)				
Threefry	106	7	1	217
Philox	107	7	0	221
(b)				
Threefry	110	4	0	224
Philox	111	2	1	222
xorshift32	87	5	22	135

(a) Results when the key was set by a call to xorshift32. (b) Results when the key or seed is incremented sequentially showing the lack of strong correlation for the counter-based generators as compared to xorshift32 alone

Merging the output streams with `interleave_streams.py` and running through dieharder gives Table 5.3a for ten million samples (Philox) and an initial xorshift32 seed of 12345.

One might complain that using xorshift32 to pick the key is also a bit of a cheat as we are close to random seeding. The claim for the counter-based generators is that even nearby keys lead to very different sequences. To test this we make a slight change to line 21 above replacing `xorshift32()` with `i+1` so that each stream is initialized with a key that is only one different from the previous stream. We do this for both Threefry and Philox. For comparison purposes we generate and interleave streams from ten xorshift32 generators (fixing the parameters) with the same seed values. Running these through dieharder gives us Table 5.3b. Clearly, the claim is supported as the interleaved streams are not showing the effect of correlation. As expected, the xorshift32 streams are strongly correlated leading to a poor showing in dieharder. A single xorshift32 stream gave a total score of 209 (107 passed, 3 weak, 4 failed).

What about running the generators in single stream mode with skipping? Again, this is easy to do,

```
1  philox4x32_key_t philox32key = {{0,0}};
2
3  void philox32_init(uint32_t seed) {
4      philox32key.v[0] = seed;
5  }
6  uint32_t philox32(uint32_t j) {
7      philox4x32_ctr_t r;
8      philox4x32_ctr_t ctr={{0,0}};
9      ctr.v[0] = j;
10     r = philox4x32(ctr, philox32key);
11     return r.v[0];
12 }
```

```
13 void main(int argc, char *argv[]) {
14     FILE *f;
15     char fname[256];
16     uint32_t i,j, nsamp, nstreams, u, key;
17     uint32_t begin, end;
18
19     nstreams = (uint32_t)atol(argv[1]);
20     nsamp = (uint32_t)atol(argv[2]);
21     key = (uint32_t)atol(argv[3]);
22     philox32_init(key);
23     for(i=0; i<nstreams; i++) {
24         sprintf(fname,"stream_%03d.dat",i);
25         f = fopen(fname,"w");
26         begin = i*nsamp;
27         end = begin + nsamp;
28         for(j=begin; j<end; j++) {
29             u = philox32(j);
30             fwrite((void*)&u, sizeof(uint32_t), 1, f);
31         }
32         fclose(f);
33     }
34 }
```
(philox_streams_skip.c)

where we initialize the generator once fixing the key in line 22. We adjust the generator call itself so that the index is given explicitly (line 6, line 29). Note, even though the call to `philox4x32` returns four values we only use the first one. Looping over the streams (lines 23–33) sets the initial index (line 26). The ending index is then `nsamp` beyond the initial since the test in line 28 does not include the end point. Line 29 gives us the desired output value which is written to disk in line 30.

In this mode, we have taken a single version of the Philox generator and are partitioning the space into `nstreams` streams of `nsamp` values. Since the size of each stream is known we set our index so that they do not overlap but fit exactly. The Threefry version is identical but returns and writes a 64-bit unsigned integer to disk instead. See *threefry_streams_skip.c*.

If we use these programs to generate ten streams of 50 million (25 million) output values each and then interleave them and test with dieharder we get,

Generator	Passed	Weak	Failed	Score
Philox	107	6	1	218
Threefry	106	8	0	220

showing that this skipping method also works well. Note, no jump function was necessary saving us from the extra steps we needed to perform in Sect. 5.3.

The results above skipped precisely the number of samples needed to not allow the streams to overlap. What if the streams do overlap? How much overlap is too

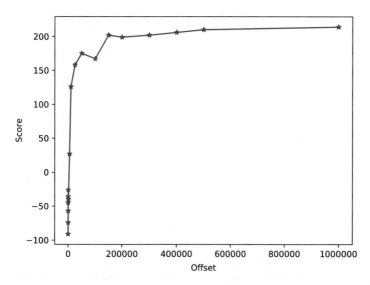

Fig. 5.3 A plot of the dieharder score as a function of the offset size between streams for the Philox generator. In each case there were ten streams of 50 million samples offset by the x-axis value. These were interleaved and tested with dieharder. When the offset reaches even 1/50-th the number of samples the resulting merged file shows good results

much? We can test this by a slight modification to the skip code above to change the starting point to be an arbitrary number of samples. Let's do this for the Philox generator so that we can build Fig. 5.3 that plots offset values (number of values skipped between streams) and the resulting dieharder score. It is clear from the figure that the Philox generator tolerates some fraction of overlap in the streams and still gives good results when the overlap is even 1/50-th of the sample size.

5.7 Discussion

In this chapter, we investigated five approaches to the problem of generating independent streams of pseudorandom values. The first is a seemingly reasonable idea: implement a pseudorandom number server. The server is accessed by processes that need pseudorandom numbers. If the server itself implements a generator with a very large period this method seems ideal. However, it has two serious drawbacks that make it unattractive. First, it is slow. Simulations often need random values at a very high rate and the overhead of interprocess communication would slow simulations considerably. Second, it is non-deterministic. There is no way to run a process that uses the server and expect the same sequence of random values when the process is run a second time. If the process really doesn't care about repeatability, this may not be an issue. After all, using the RDRAND instruction results in a non-repeatable sequence of values because of the hardware source. But, again, simulation and other models are generally quite concerned with being able to repeat a run meaning the sequence of random values needs to be repeatable as well.

Our second approach encapsulates the idea of "use a different generator per stream". This would, naturally, lead to independence. We implemented this for xorshift32 parameterizing it with 81 different sets of parameters from the original paper by Marsaglia. This leads to up to 81 different streams each using xorshfit32 but a different configuration of it. Our tests with separate xorshift32 streams showed that this approach is useful. And, since xorshift32 is a component of other generators, we saw that using this approach to create streams of KISS generators did not affect the quality of the KISS generator output. The only drawback with this approach is that not every generator can be parameterized easily and even if it can the number of parameterizations may be limited.

A common approach to parallel generators is skipping or jumping ahead. We tested this approach using MINSTD, xorshift128+ and xorshift1024*. All three of these generators can be skipped ahead. For MINSTD we skipped a single generator's output ahead by 110% of the number of samples per each stream. The xorshift128+ and xorshift1024* implementations come with a jump function pre-calculated to skip ahead by 2^{64} or 2^{512} samples, respectively. We showed that the jump function for xorshift128+ and xorshift1024* worked very well and separated the single stream into a large number of non-overlapping streams each one with a relatively large space of output values. Skipping MINSTD did not work as well but is straightforward to understand. A serious drawback to this approach is that not all generators can be skipped ahead efficiently.

Our next approach was random seeding. It is attractive because it works with any pair of generators. One simply uses a source generator's output as the seed value for a destination generator. Every time a new stream is required a new seed is used from the source generator. Our tests showed that this approach, while not without a small amount of risk, is in general quite adequate, especially if the destination generator has a large period. In that case, the probability of overlap between the streams becomes small, even vanishingly small in the case of generators like KISS.

The final approach was the Fog XOR method which is similar to random seeding but uses two separate generators whose output is XOR-ed. Each generator of each stream is initialized with a seed from a third generator. We demonstrated excellent performance even when using a very weak seed generator (RANDU).

We also demonstrated that counter-based generators can implement both parameter-based generators and skipping. Which mode is best for these generators is difficult to say though we demonstrated that subtle changes to the parameterization lead to very different sequences which is an advantage.

All of these results lead to some basic, and fairly obvious, suggestions as to how to implement generators in parallel. The best approaches are skipping, random seeding, and the Fog method. If one is satisfied with the excellent test performance of either xorshift128+ or xorshift1024* then skipping is quite attractive. A possible negative is that one needs to know how many streams have been initialized in order to call the jump function enough times to move to a new region of the output space. Random seeding, on the other hand, works well with long-period destination generators. A possible advantage here is that one need not consider how many streams have been initialized though one, naturally, needs to use the same source generator for each stream so that the seeds are successive outputs of a source generator initialized only once. The best destination generator in our tests was KISS, though we of course did not conduct exhaustive tests. The Fog method is very similar to random seeding but requires more computation. The extra layer of code may be desirable in certain applications to provide a safe-guard against poor initialization.

5.8 Chapter Summary

In this chapter we investigated five approaches for implementing pseudorandom number generators in parallel. We developed example code and tested the output of the code with the dieharder suite. We dismissed the idea of a pseudorandom number server as unattractive because it is non-deterministic limiting its use in simulation experiments. We tested the multiple generator approach by using up to 81 different versions of xorshift32. We implemented skipping that partitions the space of a single generator into non-overlapping regions for MINSTD, xorshift128+, and xorshift1024* and demonstrated good performance when interleaved streams were tested with dieharder. Random seeding involves playing the odds and simply seeding a destination generator with different seeds, one per stream, that are themselves outputs of a source generator. We saw why this approach is useful in practice and why the destination generator should have a large period. The Fog method also gives excellent results but requires implementing two generators per stream. We then investigated how to use counter-based generators for either parameterized generators or skipping and saw that either approach is viable. Finally, we discussed the results of these five approaches and made some recommendations on their use.

Exercises

5.1 One version of skipping is called "leapfrog" where a single generator outputs n consecutive samples passing one each to n streams before outputting the next n samples. Discuss the pros and cons of this approach compared to the skipping approach of Sect. 5.3.

5.2 In [3] Marsaglia presents a 64-bit version of xorshift along with a table of 275 triplets, (a, b, c), that can be used to parameterize the 64-bit generator. Write a Python version of the 64-bit generator similar to the 32-bit example in Sect. 5.2 and test its output with dieharder for 5, 10, 50, and 100 streams scaling the number of samples so that the resulting interleaved output file has 500 million samples. Is there any pattern to the test scores ($2p + w - 2f$ scores) as a function of the number of streams?

5.3 In Sect. 5.4 it was claimed that the probability of a randomly seeded generator *not* creating overlapping streams, as a function of the number of streams (s), the length of the streams (l), and the period of the generator (ρ), is $p = (1 - sl/\rho)^{s-1}$. Test this claim empirically. Fix the number of streams, $s = 10$, as well as the length of the stream, $l = 1000$, and use multiplicative linear congruential generators seeded at random. The generators to use are given below for multiplier a and modulus m. In each case, the period is $\rho = m$. These generators are from Table 2 of [7]. For each generator, count the number of times no sequences of length l, for s streams, overlap for $n = 100$ random sets of seeds. Use a respectable generator for the seeds, for example, the default Python generator (Mersenne Twister). If, for a specific generator of period ρ, there were m sets of seeds that led to no overlap for the streams then the estimated probability is $m/100$. Plot these probabilities as a function of the generator period and compare to what would be expected from the equation. The generators are,

a	m
572	16381
1944	32749
17364	65521
43165	131071
92717	262139
283741	524287
380985	1048573
360889	2097143
914334	4194301

(Hint: the MINSTD class of Sect. 5.1 might be helpful here ***)

5.4 Repeat the source and destination tests of Sect. 5.4 using (a) RANDU as source, KISS as destination and (b) KISS as source and RANDU as destination. Repeat each pairing 5–10 times to get an average dieharder score ($2p + w - 2f$). Explain the results and offer advice on pairing strong and weak generators for random seeding.

5.5 Test ten individual streams generated by skipping xorshift128+ using dieharder. Use 500 million samples per stream. Is the average score ($2p + w - 2f$) of each stream in line with the score given in Sect. 5.3?

5.6 In Sect. 5.5 we tested the Fog method using MINSTD and RANDU as the seed generators. Repeat the tests for at least three generator pairings using LCG(a,m)=LCG(33,251), LCG(572,16381), and simply incrementing the seed generator value starting at one. Explain the results. **

5.7 Repeat the Philox stream overlap test of Sect. 5.6 producing a plot like Fig. 5.3 for the MINSTD generator. Compare qualitatively with Fig. 5.3.

References

1. http://www.sprng.org/.
2. Marsaglia, George; Sullivan, Stephen. "Technical correspondence". Communications of the ACM. 36 (7) (1993): 105–110.
3. Marsaglia, George. "Xorshift rngs." Journal of Statistical Software 8.14 (2003): 1–6.
4. Kneusel, Ronald T. "Curve-Fitting on Graphics Processors Using Particle Swarm Optimization." International Journal of Computational Intelligence Systems 7, no. 2 (2014): 213–224.
5. Nguyen, Hubert. Gpu gems 3. Addison-Wesley Professional, 2007.
6. L'Ecuyer, Pierre, David Munger, Boris Oreshkin, and Richard Simard. "Random numbers for parallel computers: Requirements and methods, with emphasis on gpus." Mathematics and Computers in Simulation 135 (2017): 3–17.
7. L'ecuyer, Pierre. "Tables of linear congruential generators of different sizes and good lattice structure." Mathematics of Computation of the American Mathematical Society 68, no. 225 (1999): 249–260.
8. Fog, Agner. "Pseudo-Random Number Generators for Vector Processors and Multicore Systems." Journal of Modern Applied Statistical Methods 14, no. 1 (2015): 308–334.

Chapter 6
Cryptographically Secure Pseudorandom Number Generators

Abstract Cryptographically secure pseudorandom number generators (CSPRNGs) are pseudorandom number generators that protect against attack while still providing high quality pseudorandom values. In this chapter, we explore four of these generators, one for historical purposes (Blum Blum Shub) and three that are considered secure and are in current use: ISAAC, Fortuna, and ChaCha20.

6.1 Properties of Secure Generators

All the pseudorandom number generators we have discussed in the other chapters of this book are insecure. This means that their operation makes no attempt to prevent an attacker from learning about the state of the generator in order to predict future output or learn of earlier outputs. If the application is Monte Carlo simulation, why would we care about such things? But, simulation is not the only use for random numbers. They are also used in cryptography for, among other things,

- nonces
- keys
- one-time pads

Cryptography is a very large and very important field. A complete discussion is far beyond the scope of this book and we will make no attempt to at such here. All we will do in this chapter is present pseudorandom number generators that are considered "cryptographically secure pseudorandom number generators" (CSPRNGs) and say little to nothing about their uses in cryptography. This is grossly unsatisfying, but the topic is either its own book or something we wave our hands at and discuss very superficially. For full coverage of cryptography see [1] or any of the many excellent books on the subject.

Our working definition for a CSPRNG is:

A CSPRNG is a random number generator that passes the *next-bit* test and one where an attacker's knowledge of the state of the generator at time t makes knowing the state at any previous time impossible.

© Springer International Publishing AG, part of Springer Nature 2018   189
R. T. Kneusel, *Random Numbers and Computers*,
https://doi.org/10.1007/978-3-319-77697-2_6

The next-bit test is passed if, for bit i, there is no polynomial time algorithm operating on all previous bits, $0, \ldots, i - 1$, predicting bit i with any probability better than 50%. This is essentially the definition of a random sequence we put forward in Sect. 1.1. Yao [2] proved that any generator passing the next-bit test will pass all other (polynomial time) statistical randomness tests.

Knowledge of the state of a generator, either through an attack or a successful guess, should not compromise previously used bits. This makes the sequences of Chap. 7 unsuitable even if they are, in fact, perfectly random. If one uses $\sqrt{2}$ as a random number generator and an attacker guesses that this is the source used and that the current position in the sequence of digits of $\sqrt{2}$ is n then the attacker knows all previous and future digits. If messages were encrypted using these digits, then all previously encrypted messages would be compromised.

There have been some well-publicized examples of failures on the part of random number generators that were believed to be secure. We will examine a few of them here. The lesson from these examples is that good CSPRNGs are hard to implement properly and implementations claiming security should be left to experienced groups. Even these groups get it wrong, sometimes.

In 1996 Goldberg and Wagner detailed a random number seed weakness in the Netscape web browser [3]. The random number generator was seeded based on three values: the process id, the parent process id, time of day (seconds), and time of day (microseconds). This is for the Unix version of the web browser. Any user with an account on the system can learn the process and parent process ids. And, as they show, the time of day to one second (number of seconds since January 1st, 1970) can be easily uncovered as well. The missing microseconds can be found by brute force, only a million cases need be considered which took some 25 s on a standard microcomputer of the day. Once accomplished, the security of the secure socket layer (SSL) was compromised and encrypted browser communications could be decoded.

In [4] researchers reverse-engineered the Windows 2000 random number generator. This was, at the time, arguably the most widely used pseudorandom number generator in the world. By examining the windows generator, Dorrendorf et al. were able to determine not only the next 128 kilobytes output but the previous 128 kilobytes of the generator as well. This violates the second of the requirements above for a secure pseudorandom number generator. Microsoft later confirmed that Windows XP was similarly vulnerable.

Our last example concerns the generator of RSA keys. These depend upon $n = pq$ where p and q are unknown prime numbers but n is known. If a second key m is calculated where one of the primes from the first key is reused, say p, then $GCD(n, m) = p$ and both keys are now completely known. A large survey of such keys [5] showed that approximately 0.2% exhibited this property. In [6] Heninger et al. were able to compromise RSA and DSA private keys with a similar GCD calculation. Their analysis found that the culprit was almost always an embedded system that used poor random number seeding. For example, entropy sources, random data injected into the system generators /dev/random and /dev/urandom, were lacking during some period of time shortly after boot. In general, we will use the term *entropy* in this chapter to mean any external source that is used to alter the state of a CSPRNG. The source might be timing based on keystrokes or disk access data or network traffic data, etc. Anything that would be very difficult, if not impossible, for an attacker to know or to control completely.

In this chapter we will look at four secure generators. The first, Blum Blum Shub, is of historic interest, it should not be used for any true cryptography work. The second, ISAAC, is currently believed to be secure but does not make use of external entropy sources. Next, we will look at Fortuna by Schneier and Furguson. This generator makes use of external entropy sources and explicitly acts against state knowledge attacks (the second of our two requirements for a CSPRNG). As of this writing, Fortuna is considered secure. Finally, we consider ChaCha20 which is similar to Fortuna and is used by Google for Transport Layer Security (TLS) between its browsers and websites. It is also used in newer Linux kernels (≥ 4.8).

6.2 Blum Blum Shub

Perhaps the simplest of all secure generators is Blum Blum Shub (BBS) [7]. We are examining it here for historical reasons and because its simplicity makes it suitable for implementation directly. Plus, it is intellectually interesting as it blends an LCG-style generator (see Sect. 2.2) with prime numbers and number theory. However, to reiterate, do not use this generator for any purpose where true security is necessary.

Mathematically, BBS is straightforward. The seed is a single integer, x_0. The generator is a single statement,

$$x_{i+1} = x_i^2 \bmod n$$

mimicking the form of a linear congruential generator.

The modulus, n, is a Blum integer. A Blum integer, named after Manuel Blum (the second "Blum" in "Blum Blum Shub", his wife, Lenore, is the first), is the product of two primes, p, and q, where the primes satisfy $p \bmod 4 = 3$ and $q \bmod 4 = 3$. Primes of this form are known as Blum primes.

As $n \rightarrow \infty$, for the lowest order bits of the sequence of x_i, the lowest order bits of x_{i+1} become impossible to determine with any probability greater than 50%. From Sect. 6.1, we see that this implies satisfying the next-bit test. This is true because the problem becomes akin to solving the quadratic residuacity problem first proposed by Gauss [8]. That is, given $n = pq$, p and q both prime, and an integer, y, determine if $y = b^2 \bmod n$ for some integer, b. It is from the difficulty of this that BBS makes its claim to security. Of course, p and q must be large enough so that brute force factoring of n, known as a semiprime because its only factors are prime, cannot be accomplished in any reasonable amount of time.

In addition to the modulo 4 requirement, p and q are chosen so that $\text{GCD}(\phi(p), \phi(q))$ is small. This affects the period of the generator. Here GCD is the greatest common divisor and ϕ is Euler's totient function which for prime p is simply $p - 1$.

Let's make this practical by working with the generator. We will use Python for its intrinsic large integer support. For output we will simply take the lowest order bit of x_i. I.e. we will decide whether or not x_i is even (0) or odd (1).

Billions of prime numbers are known. We need to find two large primes that modulo 4 equal 3. From a list of two billion primes, looking at the last ten million of them, we can use a simple script to locate all the primes matching $p \bmod 4 = 3$. The two largest of these are

$$p = 47055833443$$

$$q = 47055833459$$

with $\text{GCD}(p-1, q-1) = 2$ which is small. With these two primes we now have n,

$$n = pq = (47055833443)(47055833459)$$

$$= 2214251461768250569337$$

so that the code for the generator comes together as,

```
1    import sys
2    import os
3
4    class BBS:
5        def __init__(self, seed, n=2214251461768250569337L):
6            self.seed = seed
7            self.n = n
8        def next(self):
9            b = 0
10            for i in xrange(8):
11                self.seed = (self.seed*self.seed) % self.n
12                if (self.seed % 2):
13                    b |= (1 << i)
14            return b
15
16    def main():
17        bbs = BBS(int(sys.argv[1]))
18        with open(sys.argv[3],"w") as f:
19            for i in xrange(int(sys.argv[2])):
20                f.write("%s" % chr(bbs.next()))
21
22    if (__name__ == "__main__"):
23        main()
```
(bbs.py)

where we have encapsulated the generator in a simple class, BBS. The seed value is passed to the constructor. The seed should not have p or q as a factor so pass in a number less than 47055833443. The seed is updated in line 11. We use eight such updates keeping the lowest order bit of each seed value in b. It is b that is returned so that a call to next returns a random byte. The main function simply generates a set number of bytes and dumps them to an output file.

It should be mentioned that keeping a single bit from each generator output is a very conservative approach. In [9] it was shown that more than one bit of output can be safely used and security still be maintained. In particular, the number of "safe" bits is $O(\log(\log n))$, i.e., a function of the modulus size. For our implementation this about 3.89 bits. We'll take the conservative approach and keep only one bit.

How well does this generator perform? Running the script above to generate a two billion byte output file (runtime isn't an issue here) and passing that file to ent gives,

```
Entropy = 8.000000 bits per byte.

Optimum compression would reduce the size
of this 2000000000 byte file by 0 percent.

Chi square distribution for 2000000000 samples is 254.21, and
randomly would exceed this value 50.00 percent of the times.

Arithmetic mean value of data bytes is 127.4985
     (127.5 = random).
Monte Carlo value for Pi is 3.141563067 (error 0.00 percent).
Serial correlation coefficient is -0.000046 (totally
     uncorrelated = 0.0).
```

which is about as nice a result as one could hope for. Encouraged by this, let's pass the same file to dieharder with,

```
$ dieharder -a -g 201 -f bbs.dat >dieharder_bbs.txt
```

which, when dieharder_bbs.txt is run through evaluate_dieharder_ results.py (see Sect. 4.3) gives,

```
Score: 223 for 114 tests (109 passed, 5 weak, 0 failed)
```

which is a high quality result. Let's compare this to the output of MINSTD (Sect. 2.2) which is also from the mid 1980s. In this case, we want a random output byte. The generator produces integers in the range $[0, 2^{31} - 1]$. We know from Sect. 2.2 that the low order bits of this generator are not particularly random. So, let's not keep those but instead shift the output value down 16 bits and mask with 0xFF to keep byte 3 of the output. Dumping two billion of these bytes to disk and running the file through dieharder as we did for the BBS output above gives,

```
Score: 218 for 114 tests (107 passed, 6 weak, 1 failed)
```

which is not quite as good as the BBS result.

We have not discussed the period of the BBS generator. It is a function of n and can be calculated via,

$$T = \lambda(\lambda(n))$$

where $\lambda()$ is the Carmichael λ function discussed at length in [10]. The `numbthy` Python library, available from

https://github.com/Robert-Campbell-256/Number-Theory-Python/blob/master/numbthy.py,

includes an implementation of this function. Using it gives,

$$
\begin{aligned}
T &= \lambda(\lambda(n)) \\
  &= \lambda(\lambda(2214251461768250569337)) \\
  &= 375809633466223320 \\
  &\approx 2^{58}
\end{aligned}
$$

which is a useful period but certainly not a spectacular one. Larger p and q would increase the period correspondingly. The λ function has a particularly simple form in our case,

$$
\lambda(n) = \mathrm{lcm}(\lambda(p), \lambda(q)) = \mathrm{lcm}(p - 1, q - 1)
$$

since $\lambda(p) = p - 1$ when p is prime. For composite n it is the least-common multiple (lcm) of the prime factors of n which for us are just p and q, by design.

An interesting property of the BBS generator is that it is possible to directly output the i-th value without calculating the previous values. Mathematically, x_i is,

$$
x_i = \left(x_0^{2^i \bmod \lambda(n)} \right) \bmod n
$$

We can add this to our code above for the default n value by adding two lines to the constructor,

```
self.x0 = (seed*seed) % self.n
self.l = 1107125730837069451218L
```

where `x0` holds the first value calculated when `next` is called (i.e. the next sequence value after the seed) and `l` is set to $\lambda(n) = 1107125730837069451218$. Adding a new method, then, gives us the ability to output the m-th byte in the sequence,

```
1  def idx(self, m):
2      b = 0
3      for i in xrange(8*m, 8*m+8):
4          e = pow(2, i, self.l)
5          v = pow(self.x0, e, self.n)
6          if (v % 2):
7              b |= (1 << (i%8))
8      return b
   (bbs.py)
```

where we loop eight times (line 3) to get the eight values we will extract a bit from. The bit test adds a `i%8` because the index will start on a multiple of eight. Line 4 calculates the i-th sequence value exponent. The Python `pow(a,b,c)` function quickly calculates `a**b % c` which is exactly what we need. Without this function the code would be too slow. Once we have `e` we can calculate the i-th sequence value in v (line 5). Testing this value for even or odd gives us the proper bit to put into b. Once complete, the m-th byte is returned.

A quick test shows us that the code above is working,

```
>>> from bbs import *
>>> b = BBS(1234)
>>> [b.next() for i in range(10)]
[112, 101, 8, 161, 53, 222, 129, 158, 112, 240]
>>> [b.next() for i in range(10)]
[224, 148, 248, 51, 81, 167, 151, 41, 2, 168]
>>> [b.idx(i) for i in range(10,20)]
[224, 148, 248, 51, 81, 167, 151, 41, 2, 168]
```

where we import the code and define a generator. The first list comprehension uses next to get the first ten output bytes. The second returns output bytes 10 through 19. The final list comprehension also asks for bytes 10 through 19 but by direct indexing. As we see, these byte values are the same, as expected.

The BBS generator is considered secure but it is very slow and p and q must be very large indeed to get a good period and to avoid being guessed simply. Still, the generator is intellectually interesting and simple to experiment with, especially in Python. A C implementation would likely use the GMP multiprecision integer or similar library (see https://gmplib.org/).

6.3 ISAAC

ISAAC (Indirection, Shift, Accumulate, Add, and Count) is a deterministic CSPRNG introduced by Robert Jenkins in 1994 [11]. It was released into the public domain and has withstood cryptanalysis and is, at the time of this writing, considered to be secure. In [12] Paul and Preneel contended that ISAAC is flawed but later analysis showed the implementation they used to be incorrect. Certain weaknesses to the internal state of the generator were uncovered in [13] which may compromise the first 8192 bits output. The same paper offered a fix but also commented that the overall security of ISAAC was not compromised. Finally, in [14], a plaintext attack is described which, while possible, has minimal impact on the security of ISAAC as the complexity of the attack is still very high. It is claimed that the average cycle length is about 2^{8295} with no cycle less than 2^{40} [11].

Jenkins very kindly placed ISAAC in the public domain as indicated in the reference source code found at,

http://www.burtleburtle.net/bob/rand/isaacafa.html

Therefore, we will make a version that is structured to be more in-line with the other generators we have used in this book. The core is still almost completely identical to the reference code (see readable.c on the website above).

ISAAC has an external state of 256 unsigned 32-bit integers that we refer to as the seed. It is initialized to zero but any part may be initialized prior to setting up ISAAC. As the generator runs, 256 values are created at each turn, when needed. Dividing the code into three sections we have,

```
1   uint32_t randrsl[256] = {0};
2   #define mix(a,b,c,d,e,f,g,h) \
3   { \
4       a^=b<<11; d+=a; b+=c; \
5       b^=c>>2;  e+=b; c+=d; \
6       c^=d<<8;  f+=c; d+=e; \
7       d^=e>>16; g+=d; e+=f; \
8       e^=f<<10; h+=e; f+=g; \
9       f^=g>>4;  a+=f; g+=h; \
10      g^=h<<8;  b+=g; h+=a; \
11      h^=a>>9;  c+=h; a+=b; \
12  }
```
(isaac.c)

defining the external array randrsl (keeping the reference code name). This is an array of 256 unsigned 32-bit integers representing the current set of pseudorandom values once the generator is initialized. Prior to initialization, it serves as the seed values. The mix preprocessor function is also from the reference source code.

The generator itself is a single function, isaac, that merges the initialization and execution parts of the reference code. The initialization part is,

```
13  uint32_t isaac(uint8_t flag) {
14      static uint32_t mm[256];
15      static uint32_t aa=0, bb=0, cc=0;
16      static randcnt=0;
17      uint32_t i;
18      if (flag) {
19          uint32_t a,b,c,d,e,f,g,h;
20          aa=bb=cc=0;
21          a=b=c=d=e=f=g=h=0x9e3779b9;
22          for (i=0; i<4; ++i)
23              mix(a,b,c,d,e,f,g,h);
24          for (i=0; i<256; i+=8) {
25              a+=randrsl[i  ]; b+=randrsl[i+1];
                    c+=randrsl[i+2]; d+=randrsl[i+3];
26              e+=randrsl[i+4]; f+=randrsl[i+5];
                    g+=randrsl[i+6]; h+=randrsl[i+7];
27              mix(a,b,c,d,e,f,g,h);
28              mm[i  ]=a; mm[i+1]=b; mm[i+2]=c; mm[i+3]=d;
29              mm[i+4]=e; mm[i+5]=f; mm[i+6]=g; mm[i+7]=h;
30          }
```

```
31              for (i=0; i<256; i+=8) {
32                  a+=mm[i   ]; b+=mm[i+1]; c+=mm[i+2]; d+=mm[i+3];
33                  e+=mm[i+4]; f+=mm[i+5]; g+=mm[i+6]; h+=mm[i+7];
34                  mix(a,b,c,d,e,f,g,h);
35                  mm[i   ]=a; mm[i+1]=b; mm[i+2]=c; mm[i+3]=d;
36                  mm[i+4]=e; mm[i+5]=f; mm[i+6]=g; mm[i+7]=h;
37              }
38              randcnt = 256;
39          }
```
(isaac.c)

implying that once the seed is set in randrsl calling isaac(1) will configure the generator with that seed. The code within the if-statement on line 18 is verbatim from the reference code. It scrambles the internal uint32_t variables four times (line 22) and then uses these and the seed values in randrsl, with scrambling via mix, to set up an internal state, mm. The first part of this is the loop at line 24. The second part is the loop at line 31. When the mixing of the internal state and randrsl is complete, the generator is initialized. Line 38 sets the counter to 256 so that the next part of the isaac function will calculate the first set of 256 new values like so,

```
40          if (randcnt == 256) {
41              uint32_t x,y;
42              cc = cc + 1;
43              bb = bb + cc;
44              for (i=0; i<256; ++i) {
45                  x = mm[i];
46                  switch (i%4) {
47                      case 0: aa = aa^(aa<<13); break;
48                      case 1: aa = aa^(aa>>6); break;
49                      case 2: aa = aa^(aa<<2); break;
50                      case 3: aa = aa^(aa>>16); break;
51                  }
52                  aa                = mm[(i+128)%256] + aa;
53                  mm[i]      = y    = mm[(x>>2)%256] + aa + bb;
54                  randrsl[i] = bb   = mm[(y>>10)%256] + x;
55              }
56              randcnt = 0;
57          }
58          return randrsl[randcnt++];
59      }
```
(isaac.c)

After initialization, call the generator with isaac(0) to return the next unsigned 32-bit integer. Each call increments randcnt and if all 256 values have been returned, the if-statement in line 40 executes to generate a new set of 256 values in randrsl. If randcnt is less than 256 there are unused values in the state array, simply return one and increment the counter.

Indirection, shift, accumulate, add, and count. When ISAAC is generating a new set of 256 samples (the case when randcnt is 256), it runs the loop starting on line 44 above. This loop updates the accumulator (aa) with many scrambled shifts (switch in line 46) and uses indirection by indexing into the internal state, mm, using select bits of the current element of mm which is stored in x. When ISAAC updates aa and mm and then randrsl along with bb, there is considerable mixing. Note that aa, bb, and cc are all persistent throughout the use of the generator.

Initialization of the generator is more complex than the running of ISAAC to create new output values. The seed in randrsl is used, along with the addition of continuously scrambled values like a, to initialize mm (lines 24–30). Mixing happens again and then a second pass updates mm (lines 31–37) for even more mixing of values. The slightly expanded version of [11] found on http://burtleburtle.net/bob/rand/isaac.html includes the phrase "Deducing the internal state appears to be intractable". From the mixing done above, this seems a reasonable statement.

A simple driver program will let us use ISAAC to generate an output file of unsigned 32-bit integers that we can pass to ent and dieharder to see how the generator performs. A 500 million output file (two billion bytes) using the default seed of all zeros, which is an acceptable seed value, gives ent output of,

```
Entropy = 8.000000 bits per byte.

Optimum compression would reduce the size
of this 2000000000 byte file by 0 percent.

Chi square distribution for 2000000000 samples is 287.58, and
randomly would exceed this value 10.00 percent of the times.

Arithmetic mean value of data bytes is 127.4969
    (127.5 = random).
Monte Carlo value for Pi is 3.141667875 (error 0.00 percent).
Serial correlation coefficient is -0.000061 (totally
    uncorrelated = 0.0).
```

This is an encouraging result. Running dieharder on the same output file gives a score of,

```
Score: 225 for 114 tests (111 passed, 3 weak, 0 failed)
```

Also a very good result.

The generator is easily embedded so that a simple driver program will let us use the TestU01 framework. The driver is,

```
1   unsigned int testu01_isaac(void) {
2       return (unsigned int)isaac(0);
3   }
4   int main(int argc, char *argv[]) {
5       unif01_Gen *gen;
6       isaac(1);
7       gen = unif01_CreateExternGenBits("ISAAC",
            testu01_isaac);
```

```
 8          switch (atoi(argv[1])) {
 9              case 0:   bbattery_SmallCrush(gen);    break;
10              case 1:   bbattery_Crush(gen);         break;
11              default:  bbattery_BigCrush(gen);      break;
12          }
13          unif01_DeleteExternGenBits(gen);
14          return 0;
15      }
```
(isaac.c)

where line 6 initializes ISAAC and `testu01_isaac` is a simple wrapper to return the next unsigned 32-bit integer as required by TestU01. After compiling and running we see that ISAAC passes *all* SmallCrush, Crush, and BigCrush tests. Very encouraging indeed. So, we have some confidence that ISAAC is very close to meeting the first of our two criteria for a cryptographically secure generator. We also have confidence based on the work in [14] and [13] that ISAAC is able to withstand attacks.

Given how quickly ISAAC runs, about 21.8 million unsigned integers per second on our simple test machine, ISAAC is suitable for general use. This raises the question of how well ISAAC performs with nearby seeds. If highly divergent like Threefry or Philox (see Sect. 2.6) then ISAAC might be a good candidate for parallel multistream use. With the framework developed in Chap. 5 it is straightforward to test this by generating multiple streams with a slightly different seed for each stream. We will try two approaches. The first assigns the stream number to `randrsl[0]` and leaves all other values zero. The second sets the i-th value of `randrsl` to one and leaves all the other zero, i.e., `randrsl[i]=1` where `i` is the stream number.

Following Chap. 5 we first generate ten sets of 50 million samples, repeated for each seeding approach. These are then merged into a single output file of 500 million samples using `interleave_streams.py`. The resulting two billion byte file is then passed to dieharder.

When setting the first element of `randrsl` to the stream number, $0 \ldots 9$, dieharder gives a score of,

```
Score: 222 for 114 tests (108 passed, 6 weak, 0 failed)
```

and when setting successive elements of `randrsl` to one we get,

```
Score: 220 for 114 tests (109 passed, 4 weak, 1 failed)
```

which is still a good result but one test did fail. The failing test is the "RGB lagged sum" test for a lag of 19. This test sums the samples skipping the lag number (19) before the next sample is added. A truly random result would produce a mean value of 0.5 as a floating-point number (i.e. 0xFFFFFFFF/2). The result of this sum is turned into a p-value and this p-value, the probability that a truly random generator would produce this sum for a lag of 19, using a KS-test, is quite low. In this case it is 0.00000042 which is definitely a failure. If we look at the first set of dieharder results that gave us a combined score of 222, we see that six of them were weak. Five of these six were in the RGB lagged sum test. So, while performance is still very good, our interleaved streams initialized with barely different seeds are showing some correlation, though minor.

ISAAC is a quality deterministic generator. It has withstood various attacks and is still secure. It is also fast and can be used in place of a non-cryptographically secure generator, if desired. The plethora of languages in which ISAAC has been implemented is a testament to its usefulness. Implementations exist for: C/C++, Forth, Modula-2, Delphi, Java, JavaScript, PHP, C#, Haskell, Lisp, Rust, and Go, just to name a few.

6.4   Fortuna

Fortuna was developed by Schneier and Furguson in 2003 [15] to be explicitly resistant to attacks against its state. Fortuna is best employed as a system level CSPRNG that is used over a long period of time. The Fortuna algorithm consists of three main parts [16],

1. An entropy accumulator.
2. A pseudorandom number generator.
3. A seed management system.

The entropy accumulator is how Fortuna protects against an attacker. We should note here that the word "entropy" is being used in the sense that the input to Fortuna is random or semi-random external data that an attacker should not be able to gain control of. It is not entropy in a thermodynamic nor information theory sense, though it is related at some level to the latter. For example, the ent program measures the Shannon entropy of the bytes of the input file.

Fortuna uses a system of 32 pools of entropy, each 32 bytes long. At user-defined times, new entropy is added to the pools in round-robin fashion. New entropy is hashed via SHA-256 to produce a 32 byte value. The pools are used when reseeding the generator. The first pool, P_0, is used on every reseed. Later pools are used when the number of reseeds, r, divides 2^i where i is the pool number. So, P_0 is used on every reseed, P_1 on every other, etc. The idea is that even if an attacker knows the state of some of the pools even one unknown pool will eventually restore security. Reseeding happens on a short timescale. The selected pools are hashed together to create the new seed value for the generator.

Fortuna's random number generator is typically a stream cipher in counter mode. The AES-256 cipher is recommended. We have not explored such ciphers in this book but they can be used as a random number generator when the plaintext input is a counter. Fortuna uses a 128-bit counter and a key that is the current state of the generator. With this combination, Fortuna will create a set number of output bytes, 16 at a time, as this is the output of the AES-256 cipher. In theory, one could output 2^{128} or more bytes but after this the output would start repeating. Therefore, Fortuna limits output to 2^{20} bytes in any one call. After generating the requested output, Fortuna used AES-256 to create two more hashes that are used as the new key for the next call.

The seed management system simply allows for export of the state to a disk file and import of the state from a disk file (or any other buffer). When the system starts, Fortuna needs to load the last good state from an external entity. When the system goes down, or, as recommended in [15], every 10 min or so, the state is stored for later import.

How well does Fortuna do as at generating random numbers? Security is (highly) important, of course, but if the sequence does not pass randomness tests it is of no use. The AES or other stream cipher used produces good random numbers on its own but a large number them will eventually seem suspicious because the probability of certain blocks repeating in a true random sequence will become higher than the repeat cycle for the cipher. The reseeding steps help here.

Our implementation of Fortuna, which isn't fully specified in [15] thereby giving implementers some freedom, comes from the LibTomCrypt library. This is, at the time of this writing, a seemingly well thought-of library in the public domain. Assurances of security we cannot give, however.

Download via git and install with,

```
$ git clone https://github.com/libtom/libtomcrypt.git
$ make CFLAGS="-DGMP_DESC -DUSE_GMP" -f makefile.unix
$ sudo make install -f makefile.unix
```

where we are choosing to use the GMP multiprecision integer library available here: https://gmplib.org/.

LibTomCrypt provides a full set of cryptography functions. We will only look at those for generating random numbers. The Fortuna interface provides the following,

fortuna_start	Initialize the prng_state structure.
fortuna_add_entropy	Call repeatedly with entropy data.
fortuna_ready	Ready for use. Call after adding entropy.
fortuna_read	Read blocks of data from the generator.
fortuna_done	Free memory when done.
fortuna_export	Export the current state to a buffer.
fortuna_import	Import the current state from a buffer.

The start, ready, read, and done functions are familiar enough. The export, import and add entropy functions are new. We will start with the add entropy function. From the generalized algorithm given above we know there are 32 pools of entropy used to reseed the generator at different times. The fortuna_add_entropy function is how the pools are seeded. They can be seeded even after the generator has been initialized. The function is called with,

```
fortuna_add_entropy((void*)ent, sizeof(ent), &st);
```

where st is the generator structure holding the state and ent is an array of some number of bytes. The Fortuna algorithm doesn't care where the bytes in ent come from. It is up to the user to supply data that an attacker would be unable to determine or control. It might be timing information, mouse cursor motion, some hash of network traffic, or the latest headlines from CNN. It doesn't matter and it is probably best if the sources of entropy varied on different timescales.

The import and export functions are used to preserve and restore the generator state. One reason to do this might be between shutdown and subsequent startup of the system. The primitive Fortuna functions use an external memory buffer. The user must do something intelligent with that buffer. This process is akin to stashing the 32-bit seed value of the MINSTD generator somewhere and later restoring it.

Let's see Fortuna in action and measure how well it does at generating random numbers. Because of the elegant design of the library we are using, the test code is straightforward,

```
1   #include <stdio.h>
2   #include <stdlib.h>
3   #include <stdint.h>
4   #include "tomcrypt.h"
5
6   #define N 1000000
7   void main(int argc, char *argv[]) {
8       prng_state st;
9       uint32_t n,i;
10      uint8_t *out;
11      FILE *g;
12      uint8_t ent[32] = {
13          12,53,75,63, 4,57, 8,78,89,67, 8,45,62, 3,41, 2,
14          4,56,56,67,97,88, 9,78, 5,67,45,23,12,43, 4,76};
15
16      n = (uint32_t)atol(argv[1]);
17      out = (uint8_t *)malloc(N*sizeof(uint8_t));
18      fortuna_start(}&st);
19      fortuna_add_entropy((void*)ent, sizeof(ent), }&st);
20      fortuna_ready(}&st);
21      g = fopen(argv[2],"w");
22      for(i=0; i<n; i++) {
23          fortuna_read((void*)out, N*sizeof(uint8_t), }&st);
24          fwrite((void*)out, N*sizeof(uint8_t), 1, g);
25      }
26      fclose(g);
27  }
```
(test_fortuna.c)

where we include the library headers (line 4) and define an initial entropy vector (line 12) of 32 bytes. This serves as our seed.

Line 18 initializes the state structure defined in line 8. This comes from the library and is common to all the CSPRNGs the library supports. Line 17 defines an output buffer large enough to hold N bytes. The code is set up to output nN bytes where

n is given on the command line (line 16). The newly initialized generator is seeded (line 19) and made ready. The loop in line 22 outputs n sets of N bytes. Each set is generated by the call on line 23. Note, Fortuna works faster when outputting larger blocks of data at a time. Line 24 dumps the set to disk.

Running the above for $n = 2000$ will create a two billion byte output file. This is large enough for dieharder to use and is the standard file size we have used consistently in this book. First, though, let's see what ent makes of it,

```
Entropy = 8.000000 bits per byte.

Optimum compression would reduce the size
of this 2000000000 byte file by 0 percent.

Chi square distribution for 2000000000 samples is 289.36, and
randomly would exceed this value 10.00 percent of the times.

Arithmetic mean value of data bytes is 127.4997
    (127.5 = random).
Monte Carlo value for Pi is 3.141585375 (error 0.00 percent).
Serial correlation coefficient is 0.000008 (totally
    uncorrelated = 0.0).
```

which is a solid result. The entropy is as high as it can be. This implies that the file cannot be compressed, which ent confirms. The Monte Carlo estimate of π is also accurate to nearly five digits, also a good result. The serial correlation is very low as well.

If we run dieharder against this file,

```
$ dieharder -a -g 201 -f fortuna.dat
```

and score it with our scoring metric (see evaluate_dieharder_results.py) we get,

```
Score: 226 for 114 tests (112 passed, 2 weak, 0 failed)
```

which is nearly perfect. Clearly, Fortuna is a solid pseudorandom number generator. It is also solidly secure when fed entropy that is uncompromised.

Is Fortuna deterministic? Yes, if the added entropy is deterministic. This means one could use Fortuna for simulation but that is not what it is designed to do. The entropy added must not be deterministic to ensure security.

Fortuna is, if the implementation is vetted and trusted, a good choice for secure pseudorandom numbers.

6.5 ChaCha20

Another new CSPRNG is ChaCha20 [17]. It is supported by the LibTomCrypt library, see Sect. 6.4 for installation instructions. Unlike Fortuna, ChaCha20 does not maintain entropy pools but can be reseeded after its initial seeding. Therefore,

ChaCha20 might be best used for shorter lived tasks though newer Linux kernels use it for the /dev/urandom device.

ChaCha20, like its older cousin, Salsa20, uses a complex mixing scheme built from basic XOR and modulo operations combined with rotates. This adds security in that update operations are a series of tight microprocessor instructions. It maintains a state of 512 bits and input keys of 256 bits. The "20" in "ChaCha20" comes from the number of rounds of the basic operations performed for each call. The LibTomCrypt interface is identical to that for Fortuna (Sect. 6.4),

start	Initialize the prng_state structure.
add_entropy	Call repeatedly with entropy data.
ready	Ready for use. Call after adding entropy.
read	Read blocks of data from the generator.
done	Free memory when done.
export	Export the current state to a buffer.
import	Import the current state from a buffer.

where each function name is prefixed with chacha20_prng_. All LibTomCrypt generators follow this API form and use the common state structure prng_state.

Calls to chacha20_prng_add_entropy should pass in at least 512 bits to set the state. If desired, the generator can be used as any other generator after initialization. In this sense, ChaCha20 is similar to ISAAC (Sect. 6.3). Initialize the generator by calling chacha20_prng_start, seed as above and then call chacha20_prng_ready to make the generator ready to use. Then, call chacha20_prng_read to acquire new samples and reseed, if desired, by adding more entropy later on. For completeness the state can be imported and exported.

A ChaCha20 test file looks very similar to the code used in Sect. 6.4,

```
1   #include <stdio.h>
2   #include <stdlib.h>
3   #include <stdint.h>
4   #include "tomcrypt.h"
5
6   #define N 1000000
7   void main(int argc, char *argv[]) {
8       prng_state st;
9       uint32_t n,i;
10      uint8_t *out;
11      FILE *g;
12      uint8_t ent[64] = {
13          12,53,75,63,4,57,8,78,89,67,8,45,62,3,41,2,
14           4,56,56,67,97,88,9,78,5,67,45,23,12,43,4,76,
15          74,3,8,5,34,23,134,56,6,209,45,233,2,42,33,5,
16          67,84,98,209,41,23,78,187,36,49,58,75,67,59,83,57};
17
18      n = (uint32_t)atol(argv[1]);
19      out = (uint8_t *)malloc(N*sizeof(uint8_t));
```

```
20        chacha20_prng_start(}&st);
21        chacha20_prng_add_entropy((void*)ent,
              sizeof(ent), }&st);
22        chacha20_prng_ready(}&st);
23        g = fopen(argv[2],"w");
24        for(i=0; i<n; i++) {
25            chacha20_prng_read((void*)out,
                  N*sizeof(uint8_t), }&st);
26            fwrite((void*)out, N*sizeof(uint8_t), 1, g);
27        }
28        fclose(g);
29  }
```
(test_chacha20.c)

where the output (line 26) comes from the call in line 25. The generator is initialized in lines 20–22. The 512 bits of entropy in `ent` are hard-coded in this case.

Running with $n = 2000$ to produce two billion output bytes gives us an ent output of,

```
Entropy = 8.000000 bits per byte.

Optimum compression would reduce the size
of this 2000000000 byte file by 0 percent.

Chi square distribution for 2000000000 samples is 283.82, and
randomly would exceed this value 25.00 percent of the times.

Arithmetic mean value of data bytes is 127.5027
    (127.5 = random).
Monte Carlo value for Pi is 3.141654447 (error 0.00 percent).
Serial correlation coefficient is 0.000001 (totally
    uncorrelated = 0.0).
```

which is just as good a result as Fortuna gave in Sect. 6.4. Dieharder reports,

```
Score: 221 for 114 tests (110 passed, 3 weak, 1 failed)
```

which includes a single failing test, the lagged sum test for a lag of 27 bytes. This test sums the file at intervals (the lag) and looks to see if the mean value is what would be expected for random data. This might indicate a longer-term correlation of the values generated by ChaCha20, or it might simply be one of the few random failures expected when running good random data through a statistical test suite like dieharder. To test this, at least superficially, we can alter the seed value and run through dieharder again. The seed used above was "random" in that the author simply typed on the keyboard and inserted commas to get the values. Humans are bad at random behavior, see Chap. 1. So, perhaps the failure is due to poor initialization.

Repeating the test with the first 64 bytes generated by the Middle Weyl generator from Sect. 1.3 gives a dieharder result of,

```
Score: 223 for 114 tests (112 passed, 1 weak, 1 failed)
```

where the failure is again in the RGB lagged sum test for a lag of 31 values. The one weak test was also a lagged sum test. So, with good random initialization ChaCha20 still has a dieharder failure. Exercise 6.5 asks the reader to explore this in more depth.

6.6 Chapter Summary

In this chapter we took a quick look at cryptographically secure pseudorandom number generators. These are generators that are considered secure against attacks that attempt to uncover their state to compromise their operation. We looked at Blum Blum Shub, which is of historic importance and easy to understand, but is not safe for actual, practical use. We then investigated ISAAC which is considered secure and saw that it is an excellent generator. One of the current standards is Fortuna, which we examined next. This generator goes to great lengths to secure itself from attacks and it also produces superb quality output. Next, we looked at ChaCha20, a still newer generator, that is finding increasing use for certain tasks.

Exercises

6.1 The Blum Blum Shub generator of Sect. 6.2 uses $x_{i+1} = x_i^2 \mod n$ to move to the next output value. Explore what happens to the quality of the generator if the exponent is changed to some value greater than two.

6.2 Section 6.3 tested the ISAAC generator for multiple streams by setting the initial seed element to the stream number. It was also tested by setting the element number of the stream to one. The latter caused a failure with dieharder in the RGB lagged sum test. Repeat the test for the case when the element number of the stream is set to the stream number. Does the failure persist? Does it move to a different lag value? What happens when ten other positions of the seed array are set to one instead of the first ten?

6.3 The ISAAC generator has a few "magic" numbers such as `0x9e3779b9`, the initial value of the set of unsigned integers scrambled by the `mix` macro. In the reference implementation there is a comment that `0x9e3779b9` is the golden ratio. Why? (Hint: $\phi = (1+\sqrt{5})/2$, 2^{32}, and it is really the reciprocal of ϕ). What happens to the quality of the generator if this value is altered? What happens if the shifts of `aa` are changed from 13, 6, 2, 16 to other values?

6.4 The LibTomCrypt library uses 32 entropy pools for Fortuna. This number can be set as low as four pools by changing the value of `LTC_FORTUNA_POOLS` defined in `/src/headers/tomcrypt_custom.h`. Change this value to 4 and rebuild the library. Re-run the test of Sect. 6.4 for the given seed and for multiple new seeds. For a large two billion byte file, is there any noticeable difference in the ent output?

What about dieharder output? ** (Important: when done, set the value back to 32 and build the library again!)

6.5 In Sect. 6.5 we saw that for two seedings of the ChaCha20 generator, one poor and the other random, that dieharder still reported failure for the lagged sum test. Test this more thoroughly using other quality initializations, even true random numbers from a source like HotBits, to see if this is a persistent weakness in ChaCha20 or simply bad luck for the two seeds tested.

6.6 The LibTomCrypt library used Sects. 6.4 and 6.5 supports an additional CSPRNG called "SOBER-128". Use a test program similar to those for Fortuna and ChaCha20 to evaluate this generator with ent and dieharder. Then search the literature and make a decision on whether or not this generator should be used for cryptographic purposes.

References

1. Schneier, Bruce. Applied cryptography: protocols, algorithms, and source code in C. john wiley & sons, 2015.
2. Andrew Chi-Chih Yao. Theory and applications of trapdoor functions. In Proceedings of the 23rd IEEE Symposium on Foundations of Computer Science, 1982.
3. Goldberg, Ian, and David Wagner. "Randomness and the Netscape browser." Dr Dobb's Journal-Software Tools for the Professional Programmer 21, no. 1 (1996): 66–71.
4. Dorrendorf, Leo, Zvi Gutterman, and Benny Pinkas. "Cryptanalysis of the random number generator of the windows operating system." ACM Transactions on Information and System Security (TISSEC) 13, no. 1 (2009): 10.
5. Lenstra, Arjen, James P. Hughes, Maxime Augier, Joppe Willem Bos, Thorsten Kleinjung, and Christophe Wachter. "Ron was wrong, Whit is right." (2012).
6. Heninger, Nadia, Zakir Durumeric, Eric Wustrow, and J. Alex Halderman. "Mining Your Ps and Qs: Detection of Widespread Weak Keys in Network Devices." In USENIX Security Symposium, vol. 8. 2012.
7. Blum, Lenore, Manuel Blum, and Mike Shub. "A simple unpredictable pseudo-random number generator." SIAM Journal on computing 15, no. 2 (1986): 364–383.
8. Gauss, Carl Friedrich. Disquisitiones arithmeticae. Vol. 157. Yale University Press, 1966.
9. Sidorenko, Andrey, and Berry Schoenmakers. "Concrete security of the blum-blum-shub pseudorandom generator." Lecture notes in computer science 3796 (2005): 355.
10. Erdös, Paul, Carl Pomerance, and Eric Schmutz. "Carmichael's lambda function." Acta Arith 58, no. 4 (1991): 363–385.
11. Jenkins, Robert J. "Isaac." In International Workshop on Fast Software Encryption, pp. 41–49. Springer, Berlin, Heidelberg, 1996.
12. Paul, Souradyuti, and Bart Preneel. "On the (in) security of stream ciphers based on arrays and modular addition." In ASIACRYPT, vol. 6. 2006.
13. Aumasson, Jean-Philippe. "On the pseudo-random generator ISAAC." IACR Cryptology ePrint Archive 2006 (2006): 438.
14. Pudovkina, Marina. "A known plaintext attack on the ISAAC keystream generator." IACR Cryptology ePrint Archive 2001 (2001): 49.
15. Ferguson, Niels, and Bruce Schneier. Practical cryptography. Vol. 23. New York: Wiley, 2003.
16. McEvoy, Robert, James Curran, Paul Cotter, and Colin Murphy. "Fortuna: cryptographically secure pseudo-random number generation in software and hardware." (2006): 457–Z462.
17. Bernstein, Daniel J. "The ChaCha family of stream ciphers." DJ Bernstein's Webpage. https://cr.yp.to/chacha.html.

Chapter 7
Other Random Sequences

Abstract Sequences of numbers that can be used as-is or transformed into random numbers are the topic of this chapter. We are not concerned here with the practicality of such sequences for use where pseudorandom generators are typically used, but instead present them for fun as they are interesting in their own right. We will look at the randomness of the digits in base 10 and base 16 expansions of numbers that are known to be normal or believed to be normal. We will consider the sequence of digits in factorials of different sizes. We will look at the sequence of bits generated by 1-D cellular automata, in particular Wolfram's "Rule 30". Lastly, we will consider how to use 1-D chaotic maps to generate a sequence of random numbers. In all cases we will test the sequences using either dieharder, TestU01, or ent. We conclude the chapter with a section on using genetic programming to evolve a simple pseudorandom number generator.

7.1 Introduction

The previous chapters of this book were concerned with pseudorandom number generators by which we mean algorithms that generate sequences of numbers that pass randomness tests. In general, these sequences are parameterized to some degree so that we can configure the initial state of the generator and transform that state to output the sequence of numbers.

In this chapter we will, primarily for fun and because they are intellectually interesting, consider other sequences that pass randomness tests but which are generally not suitable for use where pseudorandom generators are typically used even though a computer program can be used to generate the sequence. For example, the sequence of digits of a normal number in some base is fixed even if it passes randomness tests. The same is true for the digits of large factorials. Therefore, while one could use these sequences for the run of some system that requires random numbers, the run would be restricted to the output of just that sequence. For example, one could not test the consistency of the output by varying the starting seed and running again. Cellular automata and 1-D chaotic maps are closer to a traditional pseudorandom generator in that they can be adjusted to some degree but they still are not aligned, conceptually, with the algorithms we have explored elsewhere in this book, so we put them here.

To be specific, then, in this chapter we will consider four different kinds of sequences of numbers that exhibit randomness:

1. The expansion of normal numbers in some base.
2. Factorials of different sizes.
3. One-dimensional cellular automata.
4. One-dimensional chaotic maps.

In each case we will discuss the origin of the sequence and apply the randomness tests of Chap. 4, be it dieharder, TestU01, or ent, to evaluate the degree of randomness. We conclude the chapter with a section using genetic programming to evolve a simple 16-bit pseudorandom number generator.

7.2   Using Normal Numbers

In mathematics, a number is considered normal in base b if its expansion in base b has the property where all occurrences of n digits are equally likely. For example, if a number is normal in base ten then for $d \in \{0, 1, \ldots, 9\}$, d occurs equally often for any digit d. This is also true for $d_0 d_1$ and $d_0 d_1 d_2$ and $d_0 d_1 \ldots d_n$ for *any* possible set of n digits where n is unbounded.

While the above property seems fantastic at first it is known [1] that almost all real numbers are, in fact, normal. However, it is also true that only a few numbers have been proven to be normal. It is believed that $\sqrt{2}$, e, and π are normal but this remains unproven. For our purposes, we will take advantage of the fact that most real numbers are normal and dispense with the (very important) step of proving them to be so and instead rely upon the results of our randomness tests.

Let's start with the digits of π as this number is the most studied and literally trillions of digits have been calculated in both base 10 and base 16. We need a source of the digits of π. Fortunately, we can download the first 1 billion digits of π in hexadecimal here,

https://archive.org/details/pi_hex_1b

Expanding this archive gives us a text file with the digits in base 16. In order for us to test the digits for randomness we need to convert this file to binary. We chose the hexadecimal version for simplicity as we can simply take every two digits, treat them together as a hexadecimal number, and output a byte with that value. A simple Python script will do the job for us,

```
 1  def main():
 2      f = open("pi_hex_1b.txt","r")
 3      g = open("random_pi.dat","w")
 4
 5      f.read(2)
 6
 7      for i in xrange(1000000000):
 8          c = chr(int(f.read(2),16))
 9          g.write(c)
10
11  main()
      (pi_hex_to_bin.py)
```

where the loop in line 7 is fixed since we know the number of digits in the file. Once this script ends we have a new file, random_pi.dat, with 500 million bytes in it. A quick test of randomness is to use the ent program. This gives us,

```
Entropy = 8.000000 bits per byte.

Optimum compression would reduce the size
of this 500000000 byte file by 0 percent.

Chi square distribution for 500000000 samples is 258.92, and
randomly would exceed this value 50.00 percent of the times.

Arithmetic mean value of data bytes is 127.4988
    (127.5 = random).
Monte Carlo value for Pi is 3.141574957 (error 0.00 percent).
Serial correlation coefficient is -0.000092 (totally
    uncorrelated = 0.0).
```

which is pretty convincing as to the probable randomness of the digits of π. Indeed, it is satisfying and a bit magical that the series of digits which make up π can be used via a Monte Carlo simulation to so well estimate π itself.

In Sect. 4.5 we developed a metric based on the results of the ent program that we can use to compare the randomness of files (see process_ent_results.py). Applying this metric gives us a score of 0.00009265 where smaller is better. This puts us in a good position relative to other known random data files and high quality generators:

	Score
RANDOM.ORG	0.0000615
π	**0.0000927**
xorshift128+	0.0001422
KISS	0.0002028
CMWC	0.0002053
xorshift128	0.0002152
xorshift1024*	0.0500003
Middle Weyl	0.0500005
xorshift32	0.1000002
Mersenne Twister	0.2000002

If we run the TestU01 small crush suite we get,

```
========= Summary results of SmallCrush =========

 Version:          TestU01 1.2.3
 Generator:        From file
 Number of statistics:  15
 Total CPU time:   00:00:17.82

 All tests were passed
```

which is again encouraging. In Sect. 4.3 we used a metric derived from the results of the standard dieharder tests to compare the quality of different pseudorandom generators. The metric is a score, $2p + w - 2f$ where p is the number of passing tests, w is the number of weak tests, and f is the number of failing test. A perfect result is a score of 228. Let's apply the same metric in this case. If we run dieharder,

```
$ dieharder -a -g 201 -f random_pi.dat
```

capture the output and pass it through `evaluate_dieharder_results.py` we get,

```
Score 216 for 114 tests (102 passed, 12 weak, 0 failed)
```

which is consistent with other random files and high quality generators.

We can download millions of digits of other irrational and transcendental numbers in hexadecimal at these sites,

http://www.numberworld.org/constants.html
https://archive.org/download/Math_Constants

which we can process in the same way to produce binary files for testing (see `make_hex_binary.py`). Note, the `archive.org` file for e appears to be corrupt.

What if the expansion is in base 10 and not base 16? All is not lost in that case. If the number is normal then the probability of any given digit being in [0, 4] is the same as it being found in [5, 9]. This suggests a simple conversion: if the digit is in

[0, 4] output a zero, otherwise output a one. If we do this, every eight digits becomes one output byte in binary. The `make_decimal_binary.py` script does just this.

Running the binary versions of these files through dieharder, each based on 1 billion decimal digits, gives, all of which give excellent results on par with the best

	Passed	Weak	Failed	Score
$\sqrt{3}$	112	2	0	226
Euler γ	111	3	0	225
ϕ	110	4	0	224
log(10)	110	4	0	224
$\sqrt{2}$	107	7	0	221

pseudorandom generators tested in Chap. 4.

Let's get a quick feel for the randomness of the first nontrivial square roots. We do this to gain some intuition on the general claim that most real numbers are indeed normal. We can generate expansions in base 10 in a number of ways (see Problem 7.1). A straightforward way is to use arbitrary precision floating-point calculations. There are a number of high performance libraries for this but here we will use a simple approach in Python. It is found in the file APFP.py which implements arbitrary precision floating-point in standard Python by taking advantage of Python's native arbitrary precision integer arithmetic. For a full explanation of this class, arbitrary precision floating-point, and number formats in general, see [2].

The script is boringly straightforward,

```
 1  from APFP import *
 2  def main():
 3      APFP.dprec = 320000
 4
 5      with open("decimal_sqrt5.txt","w") as f:
 6          f.write(str(APFP("5").sqrt()))
 7      with open("decimal_sqrt7.txt","w") as f:
 8          f.write(str(APFP("7").sqrt()))
 9      with open("decimal_sqrt10.txt","w") as f:
10          f.write(str(APFP("10").sqrt()))
11      with open("decimal_sqrt11.txt","w") as f:
12          f.write(str(APFP("11").sqrt()))
13      with open("decimal_sqrt13.txt","w") as f:
14          f.write(str(APFP("13").sqrt()))
15      with open("decimal_sqrt14.txt","w") as f:
16          f.write(str(APFP("14").sqrt()))
```

214 7 Other Random Sequences

```
17        with open("decimal_sqrt15.txt","w") as f:
18            f.write(str(APFP("15").sqrt()))
19        with open("decimal_sqrt17.txt","w") as f:
20            f.write(str(APFP("17").sqrt()))
21        with open("decimal_sqrt18.txt","w") as f:
22            f.write(str(APFP("18").sqrt()))
23
24  main()
    (APFP_digits.py)
```

where we load the APFP class and then compute values writing them to disk as a string. Line 3 sets the output precision to 320,000 digits to give us, after conversion, 40,000 bytes. This is too small of a file for dieharder or TestU01 but not for ent. The important line in each case is,

```
APFP("5").sqrt()
```

which creates an instance of the APFP class, initialized to 5, and then computes the square root using Newton's method. The APFP class allows initialization with a string to enable arbitrarily large constants to be accurately represented.

Running the binarized output through ent gives scores of,

$\sqrt{5}$	0.00389003
$\sqrt{14}$	0.00389755
$\sqrt{13}$	0.00569449
$\sqrt{15}$	0.00641251
$\sqrt{18}$	0.00814498
$\sqrt{11}$	0.01135337
$\sqrt{17}$	0.01192428
$\sqrt{7}$	0.01227975
$\sqrt{10}$	0.01910024

comparing quite favorably to the results listed above, especially considering the small size of the files. Ordering the scores from smallest to largest reveals no obvious relationship between the score and the integer input.

We've demonstrated that the collection of digits in the expansion of many real numbers is suitably random. Let's now use these expansions for a qualitative simulation. In Sect. 1.4 we developed programs to create and display fractals, ifs.py and ifs_plot.py. Exercise 1.3 asks the reader to modify the IFS program to read input from a file. Using this modified ifs.py program (left as an exercise for the reader) we can create fractal images using the binary files of this section to compare, visually, how well they perform as a random sequence.

Figure 7.1 shows a zoomed in version of the IFS fern fractal using 1,000,000 points where the random numbers are from the binary files we created above.

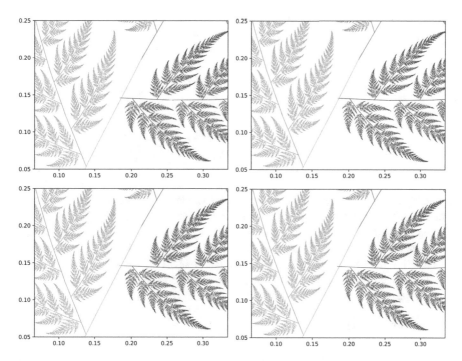

Fig. 7.1 Zoomed in fern fractals generated with IFS. In each case, 1,000,000 points were generated with random values coming from the binary files created from the hexadecimal expansions of different constants believed to be normal. The constants are, clockwise from upper left, e, $\log(10)$, ϕ, and $\sqrt{3}$

The constants are, clockwise from upper left, e, $\log(10)$, ϕ, and $\sqrt{3}$. Clearly, each of these numbers has generated a sequence that is useful in this context and leads to the same figure each time. The full fern fractal is shown in Fig. 1.5.

From the results above it is clear that were it possible to very quickly generate an arbitrary number of digits from the expansion of many, if not most, irrational numbers, we would have a powerful means for generating good random sequences. One might even combine sequences or simply use a new irrational as one currently uses a new seed for an existing pseudorandom generator.

7.3 Using Factorials

Are the digits of a large factorial, ignoring the accumulated zeros at the end, useful as a random sequence? Answering that question is the goal of this section.

To be explicit, we define the factorial of an integer $n \geq 0$ to be $n! = n(n-1)(n-2)(n-3)\ldots(2)(1)$ and further define $0! = 1$. It is clear that the factorial grows quickly,

2!	2
5!	120
7!	5040
11!	39916800
13!	6227020800
15!	1307674368000
17!	355687428096000
21!	51090942171709440000
33!	8683317618811886495518194401280000000

How can we generate large factorials, factorials with thousands, tens of thousands, and even hundreds of thousands or more digits? Fortunately it is trivially straightforward in Python because Python includes arbitrary precision integer arithmetic and factorials are integers. The classic computer science 101 definition of a factorial function is recursive. While elegant, that won't work for us because the Python stack depth will be quickly exceeded. The iterative definition is, thankfully, almost as trivial,

```
1  def fact(n):
2      if (n<2):
3          return 1
4      s = 1
5      for i in xrange(1,n+1):
6          s *= i
7      return s
   (factorials.py)
```

where we need not worry about the size of the integers involved.

One small issue needs to be addressed, however, before we investigate the randomness of large factorials. As evident in the small table above, larger and larger factorials end with more and more zeros. We don't want these. One solution is to convert the integer to a string and count backwards until we find where the zeros begin. We will turn the integer into a string, ultimately, so this approach is viable. However, it is not elegant. We would like to *know* how many zeros there will be at the end of $n!$ by looking at n before the factorial is even computed. There is an algorithm for doing just this. In code it is,

```
1  def trailing(n):
2      s = 0
3      while (n > 1):
4          s += int(n/5.0)
5          n = n/5.0
6      return s
   (factorials.py)
```

A zero is added to the end of the factorial whenever a multiple of 5 is present in n. The algorithm counts the number of 5 factors in n.

At this point we know how to generate a large integer representing $n!$ and how many zeros will be at the end of it. We need to take this large decimal and convert it into a binary file we can use with our test programs. We already discussed one approach in Sect. 7.2, namely, pass over the string representation of $n!$, digit by digit, and output a zero when the digit is in [0, 4] and a one otherwise. Putting all of this together gives us,

```
1  import sys
2  import os
3  def fact(n):
4      if (n<2):
5          return 1
6      s = 1
7      for i in xrange(1,n+1):
8          s *= i
9      return s
10
11 def trailing(n):
12     s = 0
13     while (n > 1):
14         s += int(n/5.0)
15         n = n/5.0
16     return s
17
18 def main():
19     n = int(sys.argv[1])
20     oname = sys.argv[2]
21     q = str(fact(n))[:-(trailing(n))]
22     if (len(q)%8 != 0):
23         q = q[:-(len(q)%8)]
24
25     with open(oname,"w") as f:
26         k=0
27         while (k < len(q)):
28             b = 0
29             for i in range(8):
30                 v = 1 if int(q[k])>4 else 0
31                 b |= v<<(7-i)
32                 k += 1
33             f.write("%s" % chr(b))
   (factorials.py)
```

The `main` function reads two command line arguments, n and the output filename for the binary representation (`oname`). Line 21 calculates $n!$, immediately converts it to a string, and removes the trailing zeros. Since we are converting digit to bit, we need to keep a multiple of eight, which is what lines 22 and 23 do. Finally, we can generate the output bits (lines 25–33) by looping over digits in `q` forming a byte (`b`) bit by bit (lines 30–31).

The factorial of 500,000 has 2,632,342 digits of which the trailing 124,999 are zero. This translates into an output file of 313,417 bytes. Running ent on this file gives us,

```
Entropy = 7.999373 bits per byte.

Optimum compression would reduce the size
of this 313417 byte file by 0 percent.

Chi square distribution for 313417 samples is 272.63, and
   randomly would exceed this value 25.00 percent of the times.

Arithmetic mean value of data bytes is 127.6336
   (127.5 = random).
Monte Carlo value for Pi is 3.149475458 (error 0.25 percent).
Serial correlation coefficient is -0.001209 (totally
   uncorrelated = 0.0).
```

which is reasonable for this small file and motivates us to look further. One option is to merge the output of multiple factorials. This will give us a larger output file and, hopefully, mix in some randomness, too.

Repeatedly calling factorials.py is wasteful because $(n + 1)! = (n + 1)n!$ so once we know a large factorial, say 500,000!, we can calculate 500,001! trivially as $500,001(500,000!)$. Therefore, a small modification to the bottom part of factorials.py will let us more efficiently, but still rather slowly, calculate a sequence of factorials that we can merge together into a single output file. In code,

```
 1  def main():
 2      n = int(sys.argv[1])
 3      m = int(sys.argv[2])
 4      oname = sys.argv[3]
 5
 6      for t in xrange(n,n+m):
 7          if (t == n):
 8              z = fact(t)
 9          else:
10              z = t*z
11
12          q = str(z)[:-(trailing(t))]
13          if (len(q)%8 != 0):
14              q = q[:-(len(q)%8)]
15
16          with open(oname+"_%d" % t,"w") as f:
17              k=0
18              while (k < len(q)):
19                  b = 0
20                  for i in range(8):
21                      v = 1 if int(q[k])>4 else 0
22                      b |= v<<(7-i)
23                      k += 1
24                  f.write("%s" % chr(b))
    (factorials_seq.py)
```

where we first calculate the base factorial (line 8) and then calculate the follow on factorials by multiplication by the next higher integer. If we do this for factorials from 499,991! to 500,110!, an arbitrary range, and merge them with `interleave_streams.py`, we get an output file of 37,609,440 bytes. Running ent on this gives us,

```
Entropy = 7.999996 bits per byte.

Optimum compression would reduce the size
of this 37609440 byte file by 0 percent.

Chi square distribution for 37609440 samples is 229.62, and
    randomly would exceed this value 75.00 percent of the times.

Arithmetic mean value of data bytes is 127.4932
    (127.5 = random).
Monte Carlo value for Pi is 3.141752071 (error 0.01 percent).
Serial correlation coefficient is 0.000359 (totally
    uncorrelated = 0.0).
```

which is significantly better than 500,000! alone.

The output file is rather small compared to other files we have worked with in this chapter but passing the file to dieharder and processing the dieharder output gives,

```
Score: 108 for 114 tests (72 passed, 16 weak, 26 failed)
```

which is, possibly, due to the small size of the file. The SmallCrush TestU01 battery of tests is also similarly unhappy with the small file size,

```
========= Summary results of SmallCrush =========

Version:        TestU01 1.2.3
Generator:      From file
Number of statistics:   15
Total CPU time:    00:00:17.62
The following tests gave p-values outside [0.001, 0.9990]:
(eps   means a value < 1.0e-300):
(eps1 means a value < 1.0e-15):

        Test                        p-value
-------------------------------------------------
  1  BirthdaySpacings                  eps
  2  Collision                         eps
  3  Gap                               eps
  5  CouponCollector                8.9e-10
  6  MaxOft                            eps
  9  HammingIndep                      eps
-------------------------------------------------
All other tests were passed

input file rewound 24 times
```

Note that the file was rewound 24 times. Also note that the failing tests were some of the basic tests developed in Chap. 4.

To see if these poor results are valid or the result of a small input file, let's run the file through our basic randomness tests from Chap. 4. The output of these tests is,

```
gap test:
    runs below the mean, chisq = 16.22766149,
        p-value = 0.01348688
    runs above the mean, chisq = 20.00923108,
        p-value = 0.06439448

serial test:
    chisq  = 431.55481815, p-value = 0.02728445

chisq (frequency) test:
    chisq  = 257.86424004, p-value = 0.56192677

max-of-t test:
    max-of-5 chisq  = 22.19424485, p-value = 0.72530408

correlation test:
    corr = 0.00068 (n=9402360), expected 95% CI=[-0.00065,
        0.00065], test FAILED

permutation 3 test:
    permutation size=3, chisq  = 2.86127653,
        p-value = 0.17395410

excursions test:
    v=-4, chisq=3.2514, p=0.3387166
    v=-3, chisq=6.1752, p=0.7104585
    v=-2, chisq=3.8682, p=0.4314492
    v=-1, chisq=8.5668, p=0.8723582
    v= 1, chisq=7.2266, p=0.7956726
    v= 2, chisq=6.0689, p=0.7004271
    v= 3, chisq=10.7983, p=0.9444716
    v= 4, chisq=4.6738, p=0.5430329
```

How to interpret these results? The only clear failures by our strict standards of Chap. 4 are runs below the mean, the serial test, and possibly the correlation test. However, we also learned in Chap. 4 that the gap test is unreliable for files with less than 150 MB or so. This file is about 37 MB, so we should ignore those results. Additionally, the correlation test returned failure but barely so. The ent test would be happy with a correlation of 0.00068. Also, neither TestU01 nor dieharder would consider a χ^2 p-value of 0.027 to be a failure.

With the considerations above, it seems safe to say that the merged factorial values are tending towards random, but the number of large factorials necessary to build the small data file makes it difficult to see any practical use for the approach.

7.4   Using Cellular Automata

Cellular automata are, for our purposes, discrete arrays of "cells" each of which is in one of a finite number of states. A rule is applied to an initial configuration of cells to move it to a new configuration. This is one discrete time step. The rule is applied to each cell to determine the state of the output cell. The process then repeats.

Cellular automata were initially discovered in the 1940s by Ulam and Von Neumann but became a popular computer science exercise after Conway published his "Game of Life" in 1970 [4]. The Game of Life is a 2D cellular automata with many fascinating properties. We will, however, restrict ourselves here to 1D cellular automata with two states per cell: alive (1) or dead (0). Wolfram refers to these as elementary cellular automata [5].

An example will clarify things. We will represent "the world" as a binary vector of some length,

where cells are either alive (1) or dead (0). In order to move the world to a new state we need to apply some rule to each cell. Rules are arbitrary. For Conway's game the rules are,

1. If a living cell has fewer than two living neighbors, it dies.
2. If a living cell has more than three living neighbors, it dies.
3. If a dead cell has exactly three living neighbors, it becomes alive.

leading to very complex behaviors in the 2D space of the game. All of the cellular automata we will work with in this section consider only the current cell, the cell to its immediate right, and the cell to its immediate left. As the states of the cells are binary, for any center cell, one of eight possibilities exist,

| 111 | 110 | 101 | 100 | 011 | 010 | 001 | 000 |

so that 000 means a dead cell where both its left and right neighbors are also dead while 011 means a living cell where its right neighbor is also alive and its left neighbor is dead.

We can express rules in binary as well by indicating the output bit corresponding to the state of the three bits in the current world. For example, if the center cell and its right neighbor are alive, but the left is dead, the input is 011, and we can assign a desired output for the center cell in the next iteration as 1 or 0. Doing this for each of the eight possible input sets tells us how to apply the rule. For example, one rule might be,

111	110	101	100	011	010	001	000
0	1	0	0	0	0	1	0

where the top row is the input (state of the center cell and its immediate neighbors) and the bottom row is the new state of the center cell. Since the world is finite in size we do have to make a decision as to the state of the cells just beyond the left and right edges of the world. We will assume for now that these cells are dead (0) and remain so always.

Let's apply the rule above to a random state to produce a new state,

0	1	1	1	0	1	0	1	1	0	1	0	1	0	1	1	0	1

$$\downarrow$$

1	0	0	1	1	1	1	0	1	1	1	1	1	1	0	1	1	1

where the new state uses 0 for cells beyond the left and right edge.

We indicated the rule above by showing the output for each of the eight possible inputs. As the reader has doubtless noticed, the inputs are simply the bit positions corresponding to the bits of a single byte. As such, there are 256 possible rules expressible as bytes. The rule defined above is,

$$01000010_2 = 42_{16} = 66$$

A simple Python script will allow us to apply any rule to an initial unsigned 16-bit state and watch it evolve over time. The code is,

```
1  import sys
2  def pp(j,state):
3      print "%04d:" % j,
4      for i in range(16):
5          print "1" if ((state>>(15-i))&1) else "0",
6      print
7
8  def main():
9      rule = int(sys.argv[1],16)
10     state= int(sys.argv[2],16) & 0xFFFF
11     steps= int(sys.argv[3])
12     pp(0,state)
13     for i in xrange(steps):
14         nstate = 0
15         for j in range(16):
16             if (j == 0):
17                 if (rule & (1<<((state&3)<<1))):
18                     nstate |= 1
19             else:
20                 if (rule & (1<<((state>>(j-1))&7))):
21                     nstate |= (1<<j)
22         state = nstate
23         pp(i+1,state)
   (cellular.py)
```

with pp a simple pretty-print function to output the state in binary. After reading the rule, initial state, and number of time steps from the command line (lines 9–11) the initial state is displayed (line 12) and the main loop starts (line 13). The new state (nstate) is built by examining each bit of the current state (line 15). If the current bit is bit 0 we have a special case. Let's look at the general case first, bits 1 through 15.

Line 20 applies the rule to the current bit. This is compact code, which we will see in much the same form again below, so let's break it down into its constituent parts. First, we have the phrase `(state>>(j-1))`. This shifts the current state down by `(j-1)` bits. We use $j - 1$ instead of j because this will leave bit j as bit 1. This is what we want because we need to look at this bit and its immediate neighbors. We now have these three bits in bit positions 0, 1, and 2. Next, we keep only these bits by AND-ing with 7 (111_2). At this point we have `((state>>(j-1))&7)` which is the current state of bit j and its neighbors. We need to compare this to our rule. As the rule is defined by bit positions the value of these three bits is precisely the bit position in the rule we need to examine to decide the output bit for position j. So, we shift 1 up by this many bit positions `((1<<((state>>(j-1))&7)))` and AND with the bit in that position for the rule. The result will be either zero, the bit isn't set meaning the output bit should be zero, or some non-zero value in which case we explicitly set the proper bit in the new state with line 18. As the new state is initialized to zero we do not need to explicitly set the zero case.

The paragraph above covers all bits with permanent zero boundary conditions (shifting down 15 bits leaves a zero where we want it) except for the first bit. In that case, we keep the two lowest bits `((state&3))` and shift up one position to move the right side zero into place. We then finish comparing with the rule as before.

Updating the state (line 22) and displaying it (line 23) completes the loop for the current iteration.

Let's run `cellular.py` for the initial condition and rule from above. The command line is,

```
$ python cellular.py 0x42 0x75ab 20
```

where we need to express the rule (decimal 66) and the state in hexadecimal. We'll run 20 iterations from this initial state and display the new state at each time step,

```
0000: 0 1 1 1 0 1 0 1 1 0 1 0 1 0 1 1
0001: 1 0 0 1 0 0 0 0 1 0 0 0 0 0 0 1
0002: 0 0 1 0 0 0 0 1 0 0 0 0 0 0 1 0
0003: 0 1 0 0 0 0 1 0 0 0 0 0 0 1 0 0
0004: 1 0 0 0 0 1 0 0 0 0 0 0 1 0 0 0
0005: 0 0 0 0 1 0 0 0 0 0 0 1 0 0 0 0
0006: 0 0 0 1 0 0 0 0 0 0 1 0 0 0 0 0
0007: 0 0 1 0 0 0 0 0 0 1 0 0 0 0 0 0
0008: 0 1 0 0 0 0 0 0 1 0 0 0 0 0 0 0
0009: 1 0 0 0 0 0 0 1 0 0 0 0 0 0 0 0
0010: 0 0 0 0 0 0 1 0 0 0 0 0 0 0 0 0
0011: 0 0 0 0 0 1 0 0 0 0 0 0 0 0 0 0
0012: 0 0 0 0 1 0 0 0 0 0 0 0 0 0 0 0
0013: 0 0 0 1 0 0 0 0 0 0 0 0 0 0 0 0
0014: 0 0 1 0 0 0 0 0 0 0 0 0 0 0 0 0
0015: 0 1 0 0 0 0 0 0 0 0 0 0 0 0 0 0
0016: 1 0 0 0 0 0 0 0 0 0 0 0 0 0 0 0
0017: 0 0 0 0 0 0 0 0 0 0 0 0 0 0 0 0
0018: 0 0 0 0 0 0 0 0 0 0 0 0 0 0 0 0
0019: 0 0 0 0 0 0 0 0 0 0 0 0 0 0 0 0
0020: 0 0 0 0 0 0 0 0 0 0 0 0 0 0 0 0
```

We see that rule 66 leads to extinction after 16 iterations. In fact, rule 66 leads to extinction for any initial state, much like Order 66 led to the near extinction of the Jedi^{TM}.

The code above enforces a boundary condition where the cells beyond the leftmost and rightmost are always assumed to be zero. We can relax this by instead letting the cells wrap around. If the vector has n cells labeled 0 through $n - 1$, then when updating cell 0 we consider cells 1,0, an $n - 1$ in that order. Similarly, updating cell $n - 1$ means looking at cells 0, $n - 1$, and $n - 2$, also in that order. The modification to the code above is straightforward leading to a main loop of,

```
 1  for i in xrange(steps):
 2      nstate = 0
 3      for j in range(16):
 4          if (j == 0):
 5              if (rule & (1<<(((state&3)<<1)|((state>>15)&1))))):
 6                  nstate |= 1
 7          elif (j == 15):
 8              if (rule & (1<<(((state>>14)&3)|((state&1)<<2))))):
 9                  nstate |= (1<<15)
10          else:
11              if (rule & (1<<((state>>(j-1))&7))):
12                  nstate |= (1<<j)
13      state = nstate
14      pp(i+1,state)
    (cellular_circle.py)
```

where we now look at bits 0 and 15 as special cases. When building the index into the rule for these new cases we add in the corresponding upper (bit 0) or lower (bit 15) bit value in the proper position before indexing the rule.

If we run this code with the same rule 66 and initial starting state as above we now get,

```
0000: 0 1 1 1 0 1 0 1 1 0 1 0 1 0 1 1
0001: 0 0 0 1 0 0 0 0 1 0 0 0 0 0 0 1
0002: 0 0 1 0 0 0 0 1 0 0 0 0 0 0 1 0
0003: 0 1 0 0 0 0 1 0 0 0 0 0 0 1 0 0
0004: 1 0 0 0 0 1 0 0 0 0 0 0 1 0 0 0
0005: 0 0 0 0 1 0 0 0 0 0 0 1 0 0 0 1
0006: 0 0 0 1 0 0 0 0 0 0 1 0 0 0 1 0
0007: 0 0 1 0 0 0 0 0 0 1 0 0 0 1 0 0
0008: 0 1 0 0 0 0 0 0 1 0 0 0 1 0 0 0
0009: 1 0 0 0 0 0 0 1 0 0 0 1 0 0 0 0
0010: 0 0 0 0 0 0 1 0 0 0 1 0 0 0 0 1
0011: 0 0 0 0 0 1 0 0 0 1 0 0 0 0 1 0
0012: 0 0 0 0 1 0 0 0 1 0 0 0 0 1 0 0
0013: 0 0 0 1 0 0 0 1 0 0 0 0 1 0 0 0
0014: 0 0 1 0 0 0 1 0 0 0 0 1 0 0 0 0
0015: 0 1 0 0 0 1 0 0 0 0 1 0 0 0 0 0
```

```
0016:  1 0 0 0 1 0 0 0 0 1 0 0 0 0 0 0
0017:  0 0 0 1 0 0 0 0 1 0 0 0 0 0 0 1
0018:  0 0 1 0 0 0 0 1 0 0 0 0 0 0 1 0
0019:  0 1 0 0 0 0 1 0 0 0 0 0 0 1 0 0
0020:  1 0 0 0 0 1 0 0 0 0 0 0 1 0 0 0
```

where extinction no longer happens but the bits cycle to the left indefinitely.

All of the this is interesting, certainly, but what has any of it to do with random numbers? The link comes from an observation Wolfram made in the 1980s when exploring these elementary cellular automata in great detail. It is claimed that Rule 30 is, in fact, chaotic and can be used to generate high quality random numbers. We will now investigate this claim.

If we run Rule 30 using a single initial cell and wrapping at the boundaries we get,

```
0000:  0 0 0 0 0 0 0 1 0 0 0 0 0 0 0 0
0001:  0 0 0 0 0 0 1 1 1 0 0 0 0 0 0 0
0002:  0 0 0 0 0 1 1 0 0 1 0 0 0 0 0 0
0003:  0 0 0 0 1 1 0 1 1 1 0 0 0 0 0 0
0004:  0 0 0 1 1 0 0 1 0 0 0 1 0 0 0 0
0005:  0 0 1 1 0 1 1 1 1 0 1 1 1 0 0 0
0006:  0 1 1 0 0 1 0 0 0 0 1 0 0 1 0 0
0007:  1 1 0 1 1 1 1 0 0 1 1 1 1 1 1 0
0008:  1 0 0 1 0 0 0 1 1 1 0 0 0 0 0 0
0009:  1 1 1 1 1 0 1 1 0 0 1 0 0 0 0 1
0010:  0 0 0 0 0 0 1 0 1 1 1 1 0 0 1 1
0011:  1 0 0 0 0 1 1 0 1 0 0 0 1 1 1 0
0012:  1 1 0 0 1 1 0 0 1 1 0 1 1 0 0 0
0013:  1 0 1 1 1 0 1 1 1 0 0 1 0 1 0 1
0014:  0 0 1 0 0 0 1 0 0 1 1 1 0 1 0 1
0015:  1 1 1 1 0 1 1 1 1 0 0 0 1 0 1
0016:  0 0 0 0 0 1 0 0 0 0 1 0 1 1 0 1
0017:  1 0 0 0 1 1 1 0 0 1 1 0 1 0 0 1
0018:  0 1 0 1 1 0 0 1 1 1 0 0 1 1 1 1
0019:  0 1 0 1 0 1 1 1 0 0 1 1 1 0 0 0
0020:  1 1 0 1 0 1 0 0 1 1 1 0 0 1 0 0
```

from this command line,

```
$ python cellular_circle.py 0x1e 0x0100 20
```

which seems like an interesting starting point. So, what is random here? The state? Actually, no, the claim is that the sequence of bits in the center cell (the one initially alive) is random. Indeed, early versions of Mathematica^{TM} used Rule 30 as the random number generator.

We will want many bits, so let's rewrite the code above in C and tailor it to generate bytes that we dump to disk. This code is almost the same but we will use unsigned 64-bit integers,

```
 1  #define N 64
 2  uint64_t apply_rule(uint8_t rule, uint64_t state) {
 3      uint64_t nstate=0, j;
 4      for(j=0; j<N; j++) {
 5          switch (j) {
 6              case 0:
 7                  if (rule & (1<<(((state&3)<<1)|((state>>
                        (N-1))&1))))
 8                      nstate |= 1;
 9                  break;
10              case (N-1):
11                  if (rule & (1<<(((state>>(N-2))&3)|((state&
                        1)<<2))))
12                      nstate |= (1ULL<<(N-1));
13                  break;
14              default:
15                  if (rule & (1<<((state>>(j-1))&7)))
16                      nstate |= (1ULL<<j);
17                  break;
18          }
19      }
20      return nstate;
21  }
22
23  void main(int argc, char *argv[]) {
24      FILE *f;
25      uint8_t b, rule=30;
26      uint32_t nsamp,i,j;
27      uint64_t state=0x0000000100000000;
28
29      nsamp = (uint64_t)atol(argv[1]);
30      f = fopen(argv[2],"w");
31      for(i=0; i<nsamp; i++) {
32          b = 0;
33          for(j=0; j<8; j++) {
34              b |= ((state>>(N/2))&1)<<(7-j);
35              state = apply_rule(rule, state);
36          }
37          fwrite((void*)&b, sizeof(uint8_t), 1, f);
38      }
39      fclose(f);
40  }
    (rule30_rng.c)
```

The main loop (lines 31–38) creates a single output byte (b) which is the center
bit of the 64-bit state vector iterated eight times (lines 33–36). The apply_rule
function is identical in form to the similar code found in cellular_circle.py but
parameterized for 64-bits.

Running this program and passing the output through ent gives us,

```
Entropy = 8.000000 bits per byte.

Optimum compression would reduce the size
of this 2000000000 byte file by 0 percent.

Chi square distribution for 2000000000 samples is 223.39, and
randomly would exceed this value 90.00 percent of the times.

Arithmetic mean value of data bytes is 127.5002
     (127.5 = random).
Monte Carlo value for Pi is 3.141616935 (error 0.00 percent).
Serial correlation coefficient is -0.000014 (totally
uncorrelated = 0.0).
```

which is a solid result. Let's pass the same file through dieharder. Doing so returns,

```
Score: 225 for 114 tests (111 passed, 3 weak, 0 failed)
```

which is also a solid result.

The ultimate test, of course, is to run the code within TestU01. This set up will suffice,

```
 1 #define N 64
 2 uint64_t state=0x0000000100000000;
 3 uint8_t rule=30;
 4
 5 uint64_t apply_rule(uint8_t rule, uint64_t state) {
 6 // ...
 7 }
 8
 9 unsigned int rule30_uint32(void) {
10     uint32_t v,i,j;
11     uint8_t b;
12     v = 0;
13     for(i=0; i<4; i++) {
14         b = 0;
15         for(j=0; j<8; j++) {
16             b |= ((state>>(N/2))&1)<<(7-j);
17             state = apply_rule(rule, state);
18         }
19         v |= b<<((3-i)*8);
20     }
21     return (unsigned int)v;
22 }
23
24 int main(int argc, char *argv[]) {
25     unif01_Gen *gen;
26
27     gen = unif01_CreateExternGenBits("Rule 30", rule30
        _uint32);
```

```
28      switch (atoi(argv[1])) {
29          case 0:   bbattery_SmallCrush(gen);   break;
30          case 1:   bbattery_Crush(gen);        break;
31          default:  bbattery_BigCrush(gen);     break;
32      }
33      unif01_DeleteExternGenBits(gen);
34      return 0;
35  }
```
(testu01_rule30_rng.c)

where `apply_rule` is identical to the version in `rule30_rng.c`. The only real change in this version is that `rule30_uint32` returns a single unsigned *32-bit* integer. This requires generating four bytes sequentially for each call.

The running the code for the SmallCrush battery of tests reports,

```
========= Summary results of SmallCrush =========

Version:              TestU01 1.2.3
Generator:            Rule 30
Number of statistics:  15
Total CPU time:    01:11:06.74

All tests were passed
```

which is good. However, note the run time. The generator as implemented is significantly slower than other generators, the entire 64-bit state vector needs to be updated eight times to create a single output byte and our implementation is quite naïve since each bit position could be updated in parallel. Still, if one really were interested in using Rule 30 as a generator, a parallel hardware implementation is quite possible since only basic bit operations are required. The slowness of our implementation precludes running it through Crush let alone BigCrush. However, the SmallCrush and dieharder results are convincing enough that Wolfram's claim about Rule 30 has merit.

What about the other 255 rules? A slight modification of `rule30_rng.c` will let us specify the rule value on the command line. We leave the implementation as an exercise for the reader. If we run all the rules, [0, 255], generating a 40 million byte output file each time, we can use ent to calculate a score. In each case the initial starting configuration was the same as for Rule 30, a single on bit in the middle of the word. The top scores are,

Rule	Score
89	0.0001523691
75	0.0001523691
225	0.0002438709
169	0.0002438709
86	0.0003192051
30	0.0003192051
101	0.0003796057
45	0.0003796057
135	0.0004629253
149	0.0004629253
137	1.1949211151
...	...
255	2.2360679775

where . . . represents all the unlisted rules that scored worse than Rule 137 but better than Rule 255. We see a large jump after Rule 149 so we use that as a cutoff and move to the next more rigorous set of tests: running dieharder on the top 10 lowest ent scores. In this case we use an output file of 2 billion bytes. The dieharder results, scored by `evaluate_dieharder_results.py`, are,

Rule	Passed	Weak	Failed	Score
30	111	3	0	225
86	111	3	0	225
135	109	5	0	223
149	109	5	0	223
45	108	6	0	222
101	108	6	0	222
75	107	7	0	221
89	107	7	0	221
169	80	5	29	107
225	80	5	29	107

From the table it appears that there are a number of rules leading to possible pseudorandom number generators. In reality, the two sets of four rules are actually different versions of the same base rule. The first set is Rule 30 and those isomorphic to it: Rules 86, 135, and 149. In Wolfram's notation, Rule 30 is the canonical rule. The same is true for Rule 45 which includes Rules 75, 89, and 101.

We can see this for Rule 30 by the following. The base rule can be written as,

111	110	101	100	011	010	001	000
0	0	0	1	1	1	1	0

with the top row the bit positions and the bottom row the rule: $00011110_2 = 30$. If we instead start with Rule 86 we can, by mirror flipping of the bits of the indices, recover Rule 30 like so,

		111	110	101	100	011	010	001	000		
86	=	0	1	0	1	0	1	1	0		
					↓						
		111	011	101	001	110	010	100	000		
		0	0	0	1	1	1	1	0	=	30

where we have flipped the bits of the indices left to right and then mapped the original value to the corresponding bit pattern. Similarly, if we start with Rule 135, take the complement of the indices, do the mapping, and then complement the output bits we will again arrive at Rule 30. Recall that the complement operation changes $0 \rightarrow 1$ and $1 \rightarrow 0$,

		111	110	101	100	011	010	001	000		
135	=	1	0	0	0	0	1	1	1		
					↓						
		000	001	010	011	100	101	110	111		
		1	1	1	0	0	0	0	1		
		0	0	0	1	1	1	1	0	=	30

and, finally, if we start with Rule 149, mirror the bits and then complement, we get back to Rule 30,

		111	110	101	100	011	010	001	000		
149	=	1	0	0	1	0	1	0	1		
					↓						
		000	100	010	110	001	101	011	111		
		1	1	1	0	0	0	0	1		
		0	0	0	1	1	1	1	0	=	30

From the above it should be no surprise that Rules 30, 86, 135, and 149 are essentially equally effective as possible pseudorandom generators. The same transformations account for the second set of results starting with Rule 45.

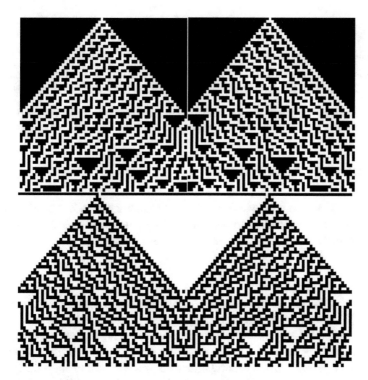

Fig. 7.2 Graphical representation of the first 64 states (top to bottom) for the rule 30 family. Bits that are set are white, unset bits are black. Clockwise from upper left: Rule 30, Rule 86, Rule 135, and Rule 149

Figure 7.2 shows the first 64 states for different rules in the Rule 30 family. The initial state is at the top. If the bit is set it is white, otherwise it is black. Clockwise from the upper left the rules are 30, 86, 135, and 149. The figure makes it easy to see what the different transformations are doing and to see why they are really the same rule. Interestingly, the dieharder scores are the same for Rule 30 and Rule 86 and, likewise, for Rule 135 and Rule 149. Visually, these pairs are only mirror images of each other.

Our test showed that the Rule 45 family also delivers good dieharder results. This is interesting because Rule 45 has not been called out as random in the way that Rule 30 has been. If we make a plot for Rule 45 along the lines of Fig. 7.2 we get Fig. 7.3 where we are showing Rule 45 for 329 steps. Note that the image has been rotated 90° to the left so that the states are now vertical and run from left to right. There seems to be some sort of order present but it may be that any structure is eliminated as the states are iterated since it appears that border effects are propagated back towards the center.

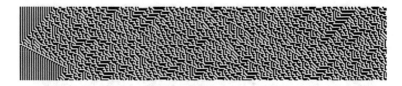

Fig. 7.3 Graphical representation of Rule 45. Bits that are set are white, unset bits are black. The image has been rotated 90° to the left so that the states are vertical and run from left to right

7.5   Using Chaotic Maps

Chaotic maps are another possible source of randomness. In this section we will investigate 1-D chaotic maps to see how we might use them to generate a random sequence. In particular, we will consider the logistic map,

$$x_{i+1} = r x_i (1 - x_i)$$

which is perhaps the best known chaotic map. For $r < 4$ the map iterates over the interval $[0, 1)$ which is already encouraging as we know well that random floating-point values in that range are ideal for processes requiring random samples.

The typical approach is to select a starting value, x_0, and iterate repeatedly for some fixed r value. The first handful of iterates are discarded to allow the system to settle into a pattern. The logistic map exhibits the period-doubling route to chaos. This means that for some r values the iteration will be fixed at a certain value, then split into two different values, then four, etc. until fully chaotic.

The standard way to illustrate this cascade is with the bifurcation plot, Fig. 7.4, showing the values of x for each r as r increases. Moving from left to right, then, reveals the period doubling route to chaos. Note, however, that even in the chaotic region there is still order.

Can we use the map as it is to generate random values? If we fix r and iterate from some starting position, x_0, ignoring an initial transient part, we will get a stream of floating-point values. These are easily turned into unsigned 32-bit integers by multiplying by 0xFFFFFFFF.

A simple C program to do just this is,

```
 1  #include <stdio.h>
 2  #include <stdint.h>
 3
 4  void main(int argc, char *argv[]) {
 5      double r,x=0.1;
 6      uint32_t i,nsamp,u;
 7      FILE *f;
 8
 9      sscanf(argv[1],"%lf",&r);
10      nsamp = (uint32_t)atol(argv[2]);
```

```
11      f = fopen(argv[3],"w");
12      for(i=0; i<1000; i++)
13          x = r*x*(1.0-x);
14      for(i=0; i<nsamp; i++) {
15          x = r*x*(1.0-x);
16          u = (uint32_t)(0xffffffff*x);
17          fwrite((void*)&u, sizeof(uint32_t), 1, f);
18      }
19      fclose(f);
20 }
```
(logistic_rng.c)

where we read the r value, number of samples, and output file name from the command line (lines 9–11). The loop in line 12 iterates the map from a fixed starting value of $x = 0.1$. This is the transient part that we throw away. Lines 14–18 generate the output samples. We iterate the map (line 15) and then make an unsigned 32-bit int (line 16) which we write to disk (line 17).

Using the ent program we can get a quick feel for how well we are doing for a given r value. For $r = 3.9$ and 1 million samples we get,

```
Entropy = 7.962741 bits per byte.

Optimum compression would reduce the size
of this 4000000 byte file by 0 percent.

Chi square distribution for 4000000 samples is 256209.32, and
randomly would exceed this value 0.01 percent of the times.
```

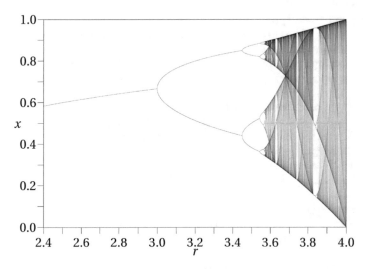

Fig. 7.4 Bifurcation plot of the logistic map, $x_{i+1} = rx_i(1 - x_i)$. The plot shows, for a particular r value, the range of x values generated by iterating the map for some starting x_0 value. When $r > 3.5699\ldots$ the map is in the chaotic region. ("Logistic Map" by Jordan Pierce, Creative Commons CC0 1.0 Universal Public Domain Dedication License)

```
Arithmetic mean value of data bytes is 133.3092
(127.5 = random).
Monte Carlo value for Pi is 2.972036972 (error 5.40 percent).
Serial correlation coefficient is -0.009653 (totally
uncorrelated = 0.0).
```

which isn't particularly good. In fact, it is terrible. Will we do better if we have a better value for r? Perhaps. Let's search for a better r value, if one exists. One way to do this is to set up a grid of r values and test each of them. A better way involves an optimization program that seeks the r value that minimizes some objective function. The objective function in our case is the score derived from the output of ent. If we capture the output of ent in a file and run process_ent_results.py against it we will get a score that we can minimize as a function of r. For example, the output above has a score of 1.00254860 for $r = 3.9$. We can constrain our search somewhat by saying that we are only interested in r values solidly in the chaotic region, [3.9, 4.0). The search itself can be conducted in many ways. Since we have no knowledge of the form of the objective function (i.e. the relationship between scores and r) we need to use a derivative-free search strategy.

Particle Swarm Optimization (PSO) [3] is a derivative-free optimization algorithm that emulates a swarm of birds. It is quite general and powerful in many different situations. Let's apply it to our search for a "best" r value. This is a one-dimensional search as our only free parameter is r and we are seeking to minimize the ent score given a particular r value.

This simple case is easily represented with the following Python code,

```
1  import sys
2  import os
3  import random
4  def evaluate(pos):
5      os.system("logistic_rng %0.10f 500000 /tmp/yyy; ent
       /tmp/yyy >/tmp/qqq" % pos)
6      os.system("python process_ent_results.py /tmp/qqq
       >/tmp/rrr")
7      try:
8          with open("/tmp/rrr") as f:
9              line = [i[:-1] for i in f.readlines()]
10             ans = float(line[1].split()[-1])
11     except:
12         ans = 1e38
13     return ans
14
15 def main():
16     imax = int(sys.argv[1])
17     npart = int(sys.argv[2])
18     wr = [0.5,0.9]
19     pos = [random.random()*0.1+3.9 for i in range(npart)]
20     vel = [0.0]*npart
```

```
21      xbest = [0.0]*npart
22      xpos = [0.0]*npart
23      gbest = 1e38
24      gpos = 0.0
25      for i in range(npart):
26          xbest[i] = evaluate(pos[i])
27          xpos[i] = pos[i]
28          if (xbest[i] < gbest):
29              gbest = xbest[i]
30              gpos = xpos[i]
31      for i in xrange(imax):
32          print "Iteration #%4d:" % (i+1),
33          w = wr[1] - (wr[1]-wr[0])*i/float(imax)
34          score = [0.0]*npart
35          for j in xrange(npart):
36              vel[j] = w*vel[j] + 1.5*random.random()*(xpos
                    [j]-pos[j]) + 1.5*random.random()*(gpos-pos[j])
37              pos[j] = pos[j] + vel[j]
38              if (pos[j] < 3.9) or (pos[j] >= 4.0):
39                  pos[j] = random.random()*0.1+3.9
40              score[j] = evaluate(pos[j])
41          for j in xrange(npart):
42              if (score[j] < xbest[j]):
43                  xbest[j] = score[j]
44                  xpos[j] = pos[j]
45              if (score[j] < gbest):
46                  gbest = score[j]
47                  gpos = pos[j]
48          print " best r=%0.10f (score=%0.10f)" % (gpos, gbest)
    (logistic_pso.py)
```

where the objective function is evaluate (lines 4–13) which itself calls
logistic_rng with the current r value (each particle represents a single r value). It
then calls ent and process_ent_results to generate a score. It is this score which
is minimized.

Lines 16–30 set up the particle swarm itself. Each particle is a single r value and
the swarm explores the space of r values ([3.9, 4.0)) comparing scores until a best
score is found (gbest) at an r value of gpos.

Particle swarm optimization moves the particles (r values) according to the
equations in lines 36 and 37. These are new candidate r values that are tested
(evaluate) and kept if one of two conditions are met. Either, this is the best
(smallest) score the current particle (j) has seen or the best the swarm has seen.
When the iterations are complete the swarm best r value is in gpos. Particle swarm
optimization works by exploiting the tension between the swarm's best position and
the current particle's best position (line 36). Lines 38–39 keep the particle in range,
[3.9, 4.0).

We run the swarm search with a command line like,

```
$ python logistic_pso.py 120 20
```

which says to use 120 iterations of a swarm with 20 particles. The results are stochastic and will vary from run to run but in general will converge to a good value if such a value exists. One run gave the following output,

```
Iteration # 120:   best r=3.9600575951 (score=1.0006950300)
```

indicating the best r value the swarm found is $r = 3.9600575951$. If we believe these results, then, calling `logistic_rng` with this r value *should* give us the most random output a single logistic map can give. Generating 100 million samples with this r value gives ent output of,

```
Entropy = 7.931500 bits per byte.

Optimum compression would reduce the size
of this 400000000 byte file by 0 percent.

Chi square distribution for 400000000 samples is 62760726.98,
and randomly would exceed this value 0.01 percent of the times.

Arithmetic mean value of data bytes is 128.4154
(127.5 = random).
Monte Carlo value for Pi is 3.035794830 (error 3.37 percent).
Serial correlation coefficient is 0.009868 (totally
uncorrelated = 0.0).
```

which is still rather poor. Clearly, a single logistic map is, by itself, not sufficiently random to be useful. However, what if we combine the output of more than one map?

In [6] Pareek et al. do just that. They run two chaotic tent maps, x_{i+1} and y_{i+1}, in cross-over fashion and combine their output to generate a stream of bits. If $x_{i+1} < y_{i+1}$ the output bit is 0, otherwise it is 1. Doing this for the logistic map gives us,

```
1  #include <stdio.h>
2  #include <stdint.h>
3  void main(int argc, char *argv[]) {
4      double r,x=0.1,y=0.1,xt,yt;
5      uint8_t b;
6      uint32_t i,j,nsamp;
7      FILE *f;
8
9      sscanf(argv[1],"%lf",&r);
10     sscanf(argv[2],"%lf",&x);
11     sscanf(argv[3],"%lf",&y);
12     nsamp = (uint32_t)atol(argv[4]);
13     f = fopen(argv[5],"w");
14     for(i=0; i<1000; i++) {
15         x = r*x*(1.0-x);
```

```
16              y = r*y*(1.0-y);
17          }
18          for(i=0; i<nsamp; i++) {
19              b = 0;
20              for(j=0; j<8; j++) {
21                  xt = r*x*(1.0-x);
22                  yt = r*y*(1.0-y);
23                  x = yt;
24                  y = xt;
25                  b |= ((xt>=yt)<<j);
26              }
27              fwrite((void*)&b, sizeof(uint8_t), 1, f);
28          }
29          fclose(f);
30      }
        (logistic_two_rng.c)
```

where we get r, x_0, y_0, and the number of output samples (bytes) from the command line (lines 9–12). The loop of lines 14–17 simply run the two logistic maps ($x_0 \neq y_0$) past their transient parts. The main loop starts on line 18. The loop of line 20 generates eight bits that we place in b. Note that we cross the output of the two maps (lines 23 and 24) to mix them as was done in [6]. Line 25 sets the proper bit of b. When the loop completes, b is written to disk and the outer loop cycles until all samples have been output.

If we generate 100 million samples with this program, for $r = 3.9999$, $x_0 = 0.2$, and $y_0 = 0.3$, ent gives us,

```
Entropy = 7.989358 bits per byte.

Optimum compression would reduce the size
of this 100000000 byte file by 0 percent.

Chi square distribution for 100000000 samples is 965841.81, and
randomly would exceed this value 0.01 percent of the times.

Arithmetic mean value of data bytes is 127.4816
(127.5 = random).
Monte Carlo value for Pi is 3.135217565 (error 0.20 percent).
Serial correlation coefficient is 0.005755 (totally
uncorrelated = 0.0).
```

which is better than the single logistic map gave us but still not really what we would like when compared to the digits of Sect. 7.2 or the generators of Chap. 2. Additionally, this approach runs the logistic map, a set of floating-point operations, eight times in order to create a single output byte. Perhaps we can do better still?

If we take the current xt and yt floating-point values and multiply them by 0xFFFFFFFF we will have an unsigned 32-bit integer. We can combine these two integers with some byte swapping and XOR to create an output 32-bit value. In code the relevant section is,

```
 1 | for(i=0; i<nsamp; i++) {
 2 |     xt = r*x*(1.0-x);
 3 |     yt = r*y*(1.0-y);
 4 |     x = yt;
 5 |     y = xt;
 6 |     u0 = (uint32_t)(0xFFFFFFFF*xt);
 7 |     u1 = (uint32_t)(0xFFFFFFFF*yt);
 8 |     u1 = (u1>>24)|(u1<<24)|((u1>>16)&0xff)<<8|((u1>>8)&0xff)
     |         <<16;
 9 |     u0 ^= u1;
10 |     fwrite((void*)&u0, sizeof(uint32_t), 1, f);
11 | }
   |     (logistic_three_rng.c)
```

where we cross the two map outputs as before but in addition we calculate the corresponding integers (lines 6 and 7) and then swap the bytes of the second before XOR-ing it into the first (lines 8 and 9). This entire 32-bit value is then written to disk before the next sample is computed. Note that this approach generates 32-bits at a time instead of a single bit per iteration. Let's christen this generator "randlog" to make it easier to refer to it.

A 100 million sample (400 million byte) file with $r = 3.9999$, $x_0 = 0.2$, and $y_0 = 0.3$ when run through ent gives,

```
Entropy = 8.000000 bits per byte.

Optimum compression would reduce the size
of this 400000000 byte file by 0 percent.

Chi square distribution for 400000000 samples is 255.33, and
randomly would exceed this value 50.00 percent of the times.

Arithmetic mean value of data bytes is 127.5016
   (127.5 = random).
Monte Carlo value for Pi is 3.141324631 (error 0.01 percent).
Serial correlation coefficient is 0.000033 (totally
uncorrelated = 0.0).
```

which is clearly encouraging. If we use the same parameters to generate a 2 billion byte file (500 million samples) and run through dieharder we get,

```
Score: 226 for 114 tests (112 passed, 2 weak, 0 failed)
```

which is one of the highest scores we've seen in this book from generator output tested with dieharder.

To see if the randlog generator passes TestU01 Crush and Big Crush tests requires embedding the generator along with calls to the TestU01 suite. This is straightforward in C,

```
 1  #include "TestU01.h"
 2  #include <stdio.h>
 3  #include <stdlib.h>
 4  #include <stdint.h>
 5  double r=3.9999, x=0.2, y=0.3;
 6
 7  unsigned int logistic_three(void) {
 8      unsigned int u0,u1;
 9      double xt = r*x*(1.0-x);
10      double yt = r*y*(1.0-y);
11      x = yt;
12      y = xt;
13      u0 = (unsigned int)(0xFFFFFFFF*xt);
14      u1 = (unsigned int)(0xFFFFFFFF*yt);
15      u1 = (u1>>24)|(u1<<24)|((u1>>16)&0xff)<<8|((u1>>8)&0xff)
        <<16;
16      u0 ^= u1;
17      return u0;
18  }
19
20  int main(int argc, char *argv[]) {
21      unif01_Gen *gen;
22      int i;
23      sscanf(argv[1],"%lf",&r);
24      sscanf(argv[2],"%lf",&x);
25      sscanf(argv[3],"%lf",&y);
26      for(i=0; i<1000; i++) {
27          x = r*x*(1.0-x);
28          y = r*y*(1.0-y);
29      }
30      gen = unif01_CreateExternGenBits("Logistic Three",
        logistic_three);
31      switch (atoi(argv[4])) {
32          case 0:   bbattery_SmallCrush(gen);   break;
33          case 1:   bbattery_Crush(gen);        break;
34          default:  bbattery_BigCrush(gen);     break;
35      }
36      unif01_DeleteExternGenBits(gen);
37      return 0;
38  }
    (testu01_logistic_three.c)
```

This version reads r, x_0, y_0, and which TestU01 suite to run from the command line. Lines 26–29 perform the initial set of iterations to remove the transient part. Line 30 sets up TestU01 to call logistic_three which takes no arguments and returns an unsigned 32-bit integer. Then, the proper battery of tests is called, the generator object is cleaned up (line 36), and the program ends. The test results are printed to the console. The `logistic_three` function returns the next 32-bit value for the generator. For the results below we consistently use $r = 3.9999$, $x_0 = 0.2$, and $y_0 = 0.3$.

Running the Crush battery of tests gives us,

```
========= Summary results of Crush =========

Version:              TestU01 1.2.3
Generator:            Logistic Three
Number of statistics:  144
Total CPU time:    00:39:23.46

All tests were passed
```

which is a perfect result. Running BigCrush, the strictest battery of tests gives,

```
========= Summary results of BigCrush =========

Version:              TestU01 1.2.3
Generator:            Logistic Three
Number of statistics:  160
Total CPU time:    04:51:57.60
The following tests gave p-values outside [0.001, 0.9990]:
(eps  means a value < 1.0e-300):
(eps1 means a value < 1.0e-15):

        Test                           p-value
   ---------------------------------------------
    1  SerialOver, r = 0               4.3e-5
  101  Run of bits, r = 0              4.6e-4
  103  AutoCor, d=1, r=0               0.9997
   ---------------------------------------------
All other tests were passed
```

which is nearly perfect.

We do not have a good understanding of the relationship between x_0 and y_0 and how these values affect the randomness of the output. Therefore, let's run a grid search for $x_0 \in \{0.1, 0.2, 0.3, 0.4\}$ and $y_0 = x_0 + \{0.1, 0.2, 0.3, 0.4\}$, which forces $x_0 < y_0$, just to limit the space a bit. The script that does this is `testu01_logistic_three_tests.py` and it calls the TestU01 code above to run the Crush battery on the output for each set of seed values.

The results of this investigation can be summarized by the number of Crush tests (144 total) which fail. This gives us Table 7.1 where the mean number of Crush failures is 2.2 ± 0.6. These are quite reasonable results and indicate that the randlog generator might not be too sensitive to initialization values, for most cases.

The randlog generator is initialized by the x_0 and y_0 seed values. Let's create multiple streams using a deterministic sequence of these initial values and test them as we tested the multistream generators of Chap. 5. Lacking any better intuition let's allow the user to specify the initial x_0 and y_0 and then simply set the seed values to a linear increase in these values. Specifically, if x_0 and y_0 are the base seeds then the seeds for the s stream are $x_0 + 0.05s$ and $y_0 + 0.05s$ with 0.05 pulled from thin air. A

Table 7.1 Number of randlog Crush tests failing (out of 144 tests) for different x_0 and y_0 initialization values. The mean $\pm$ SE = 2.2 ± 0.6

x_0	y_0	Failed
0.1	0.2	0
0.1	0.3	5
0.1	0.4	0
0.1	0.5	5
0.2	0.3	0
0.2	0.4	5
0.2	0.5	0
0.2	0.6	0
0.3	0.4	1
0.3	0.5	5
0.3	0.6	5
0.3	0.7	0
0.4	0.5	1
0.4	0.6	0
0.4	0.7	3
0.4	0.8	5

simple script will generate the streams. If we create 10 streams of 50 million samples each using $x_0 = 0.1$ and $y_0 = 0.2$ and merge them with `interleave_streams.py` we will have an output file of 2 billion bytes. Running ent on this file gives us,

```
Entropy = 8.000000 bits per byte.

Optimum compression would reduce the size
of this 2000000000 byte file by 0 percent.

Chi square distribution for 2000000000 samples is 303.93, and
randomly would exceed this value 2.50 percent of the times.

Arithmetic mean value of data bytes is 127.4970
    (127.5 = random).
Monte Carlo value for Pi is 3.141687063 (error 0.00 percent).
Serial correlation coefficient is -0.000018 (totally
uncorrelated = 0.0).
```

which is a reasonable result. Running dieharder on this file and applying our scoring technique gives,

```
Score: 220 for 114 tests (106 passed, 8 weak, 0 failed)
```

which is also a solid result.

As randlog uses chaotic maps in the chaotic region (fixing $r = 3.9999$), we should expect the butterfly effect to help us in that generators seeded with very close seed values should lead to very different outputs. To that end, let's test the generator with a series of seeds that differ by only a small amount. We will use the same multistream approach as above but change the increment between stream seeds. Specifically, let's run 10 streams of 50 million samples for each increment and then run the output through dieharder. We will fix the base seeds at $x_0 = 0.1$ and $y_0 = 0.2$ as before but this time change the increment to be $i \in \{0.0000001, 0.000001, 0.00001, 0.0001, 0.001, 0.01\}$ so that the seeds for each stream are $x_0 = 0.1 + is$, $y_0 = 0.2 + is$ with s the stream number $[0, 9]$ and i the increment value. In all cases there is an initial burn-in of 1000 iterations.

Running dieharder on each of the output files generated as a function of the increment gives good results, as anticipated,

Increment	Passed	Weak	Failed	Score
0.0100000	107	6	1	218
0.0010000	108	6	0	222
0.0001000	108	5	1	219
0.0000100	110	3	1	221
0.0000010	113	1	0	227
0.0000001	109	5	0	223

indicating that the randlog generator is potentially useful in a parallel setting.

How small can the increment be? The implementation uses IEEE 754 binary64 arithmetic so in theory the increment could be as small as about 2.2×10^{-308}, ignoring subnormal numbers. Let's push things a bit and repeat the tests for still smaller increments. Doing so gives us,

Increment	Passed	Weak	Failed	Score
10^{-8}	107	5	2	215
10^{-9}	110	4	0	224
10^{-10}	109	5	0	223
10^{-11}	29	8	77	**-88**
10^{-12}	30	5	79	**-93**
10^{-13}	30	5	79	**-93**

where we still see good results for increments as small as 10^{-10}, a particularly striking demonstration of sensitive dependence on initial conditions!

More extensive experiments might reveal parameter settings which perform better still. However, even the simple tests performed in this section indicate that generators based on chaotic maps have attractive properties.

7.6   An Experiment

And now for something completely different. The experiment of this section is an attempt to answer the question: can we use random numbers to evolve a random number generator? We put this section in this chapter because it also generates a random sequence but instead of a random sequence of numbers it is a random sequence of programs. We freely admit the tenuous nature of this association and proceed nonetheless.

Our goal is to try and evolve, through genetic programming [7], a 16-bit pseudorandom number generator with reasonable performance. We select 16-bits not because it is useful but because we (without justification) believe that in such a reduced space we stand some chance of success.

At its simplest, genetic programming generates random computer programs, runs those programs, measures their output against some fitness metric, updates the programs in some manner, generally by crossover (mixing parts of two programs, A and B, to produce a new program, C) and mutation (a random change to A to produce A'), and repeats until some performance criteria is met. Sophisticated genetic programming attempts to create programs directly in a programming language like C. We will back off a bit and instead create programs that are somewhat closer to assembly language. However, because of the operator set we will implement, the generated programs will be immediately representable in C.

Our basic algorithm is,

1. Create a population of n randomly generated computer programs (functions).
2. Run these programs to produce a sequence of output values, $x_0, x_1, x_2, \ldots$. We are hoping that this sequence will be random.
3. Measure the output sequence against some metric to decide if it is random or not.
4. Keep some fraction, $f_k \leq 0.5$, of the best performing programs. I.e. keep nf_k programs.
5. Breed these best programs to produce another nf_k programs that are made by selecting random cut points in each of the two parent programs and connecting the upper part of program A with the lower part of program B.
6. If $f_k < 0.5$, then the top performing programs plus the bred programs will total to some number less than n. Make up the difference by generating new random programs. This is in place of the usual mutation operator.
7. Repeat from Step 2 keeping the best performing program whenever a new overall best is found.
8. When all desired iterations are complete, report the best performing program as our solution.

The code for all of this is in `evolve.py` which makes use of `Func.py` as well as standard Python libraries and NumPy. We will show parts of the code below but not all of it. Please refer to the source files found on the book website.

For us, a program is a sequence of math operations. There is a single input value (the seed, x) and we have access to ten registers ($r_0, r_1, \ldots, r_9$). There are no loops. A single instruction will look like this,

```
<reg> = <op1> <op> <op2>
```

where `<reg>` is a register, r_n. Next, `<op1>` and `<op2>` are the two operands, either a register, x, or an unsigned 16-bit integer. The operation, `<op>`, is chosen from the standard set of C integer operators,

+	addition
-	subtraction
*	multiplication
/	division
%	modulo
|	bitwise OR
&	bitwise AND
^	bitwise XOR
<<	shift left
>>	shift right

where we also allow direct assignment, `<reg>` = `<op0>`, with the second operand ignored. All values are unsigned 16-bit integers.

A program, then, is a sequence of instructions. In practice, we represent a program as a vector of numbers, six to an instruction,

```
[op, op0, typ0, op1, typ1, dst]
```

for the operation (op), first operand value (op0) and type (typ0), the second operand value (op1) and type (typ1), followed by the destination register number (dst). The type values decide how to interpret the operand. If type is -1 the value is ignored, instead the input x is used. If type is 0, the value is a number. Finally, if type is $+1$ the value is a register number, $0, \ldots, 9$.

For example, this program,

```
r2 = r2 - x
r3 = r3 << (r0 % 16)
r2 = x >> (r2 % 16)
r5 = r5 / r1
r9 = 48632
```

is represented internally as,

$$[1, 2, 1, 0, -1, 2, 9, 3, 1, 0, 1, 3, 10, 0, -1, 2, 1, 2, 3, 5, 1, 1, 1, 5, 8, 48632, 0, 40315, 0, 9]$$

so that the Func class can execute the program by pulling groups of six values from the vector and interpreting them. This also makes crossover straightforward, simply break at a multiple of six. The return value from a program is which ever register is last assigned.

To make life simple, the Func class can also output a program as a valid C function. This makes working with the generated program straightforward. The example above is rendered in C as,

```
uint16_t gp_rand(uint16_t x) {
    uint16_t r0=0,r1=0,r2=0,r3=0,r4=0;
    uint16_t r5=0,r6=0,r7=0,r8=0,r9=0;
```

```
    r2 = r2 - x;
    r3 = r3 << (r0 % 16);
    r2 = x >> (r2 % 16);
    r5 = r5 / r1;
    r9 = 48632;
    return r9;
}
```

where the input is the seed. Note that the registers are initialized to zero on every call. This is also true when evaluation happens in the Python code. Also note that no effort is made to ensure that the sequence of instructions is meaningful. In this example, the constant 48632 is returned, always, none of the other instructions having any effect. When we look at evolved programs we will see how to pare them down to their essentials. This is like actual biological evolution, it works, but that doesn't mean it is always the most elegant solution.

The first step in our algorithm above is to generate a population of n programs. We will preserve the population number as evolution happens. The code to create the programs is,

```
1  def RandomProgram(self):
2      p = []
3      dlist = []
4      n  = int(0.5*self.pmax)
5      n += int(0.5*random.random()*self.pmax)
6      for i in range(n):
7          dst = int(10*random.random())
8          dlist.append(dst)
9          op  = int(11*random.random())
10         t = random.random()
11         if (t < self.xprob):
12             op0 = [0,-1]
13         elif (t < (self.xprob+self.rprob)):
14             if (dlist == []):
15                 op0 = [int(random.random()*10),+1]
16             else:
17                 op0 = [dlist[0],+1]
18                 dlist = dlist[1:]
19                 random.shuffle(dlist)
20         else:
21             op0 = [int(65536*random.random()),0]
22         t = random.random()
23         if (t < self.xprob):
24             op1 = [0,-1]
25         elif (t < (self.xprob+self.rprob)):
26             if (dlist == []):
27                 op1 = [int(random.random()*10),+1]
28             else:
29                 op1 = [dlist[0],+1]
30                 dlist = dlist[1:]
31                 random.shuffle(dlist)
```

```
32              else:
33                  op1 = [int(65536*random.random()),0]
34          p += [op]+op0+op1+[dst]
35      return p
   (evolve.py)
```

where we store the program in `p` as a list. Lines 4 and 5 select the number of instructions which is at least half the maximum allowed (`pmax`) and up to the full extent. We then loop (line 6) to create this many instructions. The destination register is chosen (line 7) and stored in a list (`dlist`) ensuring that selected registers are used at some future point. The operation is chosen in line 9. Then come two blocks of code to select the two operands. Each block uses a random number and compares it to two probabilities, `xprob` and `rprob`, set manually to 0.1 and 0.9, respectively. These are the likelihood of selecting the type of operand. If below 0.1, the operand is x, the argument. If above 0.1 but below 0.9, the operand is a register number. Finally, if above 0.9, the operand is a numeric constant. This is what lines 10–21 are doing while also paying attention to `dlist` to use registers a little more intelligently. Finally, line 34 puts the entire program together as a numeric vector and returns it.

With a starting population, we are now able to evaluate the programs and then move onto the evolution step. Evaluation of the programs is straightforward,

```
1 def RunProgram(self, v):
2     x = 12345
3     s = []
4     for i in range(self.ent):
5         s.append(Func(v,x).Eval())
6         x = s[-1]
7     return self.ScoreSequence(s)
  (evolve.py)
```

with the `Eval` method of the `Func` class applying in Python the same operations that will ultimately be called in C for the given value of x (line 5). We will not show the `Func` class here. The seed (x) is set and a predefined number of calls to the function are made passing the seed in and updating it with whatever value is returned. The `ent` member variable is fixed at 10,000 for reasons that will become clear shortly. When all values are generated, the sequence is passed to `ScoreSequence` to be evaluated. The score is the fitness of the program v and will be used to decide what to do with the program when constructing the next generation. The smaller the score, the better.

Let's pause a moment to consider what it is we need to minimize in order to arrive at a pseudorandom number generator. Clearly, we need to decide if the sequence in s is sufficiently random. Well, large parts of this book have concerned themselves with doing just that, so why don't we use some tool like dieharder, TestU01, or ent? In this case we won't be happy with the results. Recall, we are attempting to evolve

16-bit generators. No 16-bit generator will pass the sort of statistical tests found in dieharder or TestU01. We could use ent, which was, to be fully transparent, the first metric used. However, even though it has its uses, it is not what we want in this case.

So, what metric *do* we want? We want to use the sequence generated by a candidate program for something where a quality pseudorandom number generator would give us good results. Basically, we want to simulate something, and get a result for a good pseudorandom generator, so that we can compare that result with what a candidate generator gives us.

One fun way to do this is to use the Mersenne Twister in Python to create a Sierpinski triangle in a matrix. This will give us a mask where each element is either zero (no point fell there) or one (at least one point fell there). We can do the same thing with our candidate generators and use the squared error between the two matrices as our score. The smaller this is the closer the candidate generator's output is to what we would expect from a good generator.

The Sierpinski triangle is a fractal, see Sect. 1.4 and Fig. 1.5 (bottom row, middle). In Sect. 1.4 we used the iterated function system to create the fractal but the triangle can be formed in a much simpler way:

1. Define the three vertices of the triangle: (x_0, y_0), (x_1, y_1), (x_2, y_2).
2. Start with $x = x_0$ and $y = y_0$.
3. Select a random vertex, $n \in \{0, 1, 2\}$.
4. Plot $x \leftarrow (x + x_n)/2$, $y \leftarrow (y + y_n)/2$.
5. Repeat from Step 3.

Let's use this algorithm to fill in the mask matrix when we create the initial population,

```
 1  def CreateMaskArray(self):
 2      self.mask = np.zeros((80,80))
 3      px = np.array([0,40,79], dtype="uint16")
 4      py = np.array([0,79,0], dtype="uint16")
 5      x = px[0]; y = py[0];
 6      for i in xrange(self.ent):
 7          n = random.randint(0,2)
 8          x = (x+px[n])/2
 9          y = (y+py[n])/2
10          self.mask[x,y]=1
    (evolve.py)
```

This code creates a 2D matrix, mask, that is zero where points did not fall and one where they did fall. The mask itself looks like this when rendered as an image,

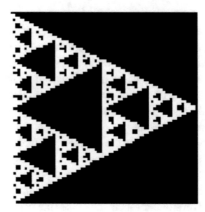

We will attempt to create a similar mask for each candidate program. The squared
error between the two masks will be the fitness score for the candidate. Specifically,

```
 1  def ScoreSequence(self,s):
 2      pmask = np.zeros((80,80))
 3      px = np.array([0,40,79], dtype="uint16")
 4      py = np.array([0,79,0], dtype="uint16")
 5      x = px[0]; y = py[0];
 6      for i in xrange(len(s)):
 7          n = int(3*(s[i]/65536.0))
 8          x = (x+px[n])/2
 9          y = (y+py[n])/2
10          pmask[x,y]=1
11      return np.sqrt(((pmask-self.mask)**2).sum())
    (evolve.py)
```

to make the mask from the sequence generated by the candidate. Line 11 calculates
the squared error between this mask and the high quality mask. We use NumPy here
to speed things along.

Once we have scores for each candidate program in the current population,
we need to decide what to do with them. We sort the scores and keep the top
nf_k programs as-is for breeding and passing to the next generation. Here n is
the population size, the number of programs, and f_k is the fraction passed on for
breeding. We then breed these programs to create another nf_k programs for the next
generation. The breeding method is straightforward,

```
 1  def Breed(self, new_pop):
 2      pop = []
 3      for i in range(len(new_pop)):
 4          p0 = random.randint(0,len(new_pop)-1)
```

```
 5              p1 = random.randint(0,len(new_pop)-1)
 6              while (p0 == p1):
 7                  p1 = random.randint(0,len(new_pop)-1)
 8              i0 = len(new_pop[p0])/6
 9              i1 = len(new_pop[p1])/6
10              b0 = random.randint(1,i0-2)
11              b1 = random.randint(1,i1-2)
12              prog = new_pop[p0][:(6*b0)] + \
13                      new_pop[p1][(6*b1):]
14              if (len(prog) > 2*6*self.pmax):
15                  prog = prog[:(2*6*self.pmax)]
16              pop.append(prog)
17          return pop
       (evolve.py)
```

where new_pop is a list of lists with each sublist a top performing program from the current generation. We want to create that many new programs by crossover (line 3). We chose two of these programs at random, never choosing the same program twice (lines 4–7). Next, we calculate the number of instructions in each program (lines 8,9) and pick two instructions at split to break them (lines 10,11). In line 12 we create the new program by simply merging the first b0 instructions of one program with the final b1 instructions of the other program. Lines 14 and 15 check to see if the new program has grown to more than twice the largest program (pmax) and if so, truncates it. This prevents programs from growing too large.

The main loop is quite short,

```
def Run(self):
    self.start = time.time()
    self.Initialize()
    self.iteration = 0
    while (not self.Done()):
        self.Msg("Iteration #%d" % (self.iteration+1))
        self.Step()
        self.iteration += 1
        if (self.verbose):
            self.Msg("")
    self.end = time.time()
```

The Initialize method sets up the initial population and builds the Sierpinski mask. The Done method simply checks to see if we have finished all the desired iterations. The Step method implements most of the steps listed above in the beginning of this section. When finished, we will have the best program found in best_prog. The full code writes several files to the given output directory including the history of the search (all best scores and associated programs), the parameters supplied, a C version of the best program (gp_rand.c), the complete console output, and a simple test program (run.c) that will create output data files using the best program. It will also attempt to measure the period of the generator.

A sample run of *evolve.py* starts with,

```
Genetic programming - evolve a 16-bit pseudorandom number
generator

Population size      : 10 programs (max len=25)
Number of generations: 10
breed top fraction   : 0.333333

Iteration #1
    evaluating and ranking current population
    mean top scores = 39.2172711936 +/- 0.0433594684
*** new best program found, score = 39.1152144312
    breeding next generation
    generating new random programs

Iteration #2
    evaluating and ranking current population
    mean top scores = 39.2002939709 +/- 0.0347335766
    breeding next generation
    generating new random programs

Iteration #3
    evaluating and ranking current population
    mean top scores = 39.0554245101 +/- 0.0488182662
*** new best program found, score = 38.9358446679
    breeding next generation
    generating new random programs
```

and ends with,

```
Iteration #10
    evaluating and ranking current population
    mean top scores = 10.8166538264 +/- 0.0000000000
    diversity too low, randomizing
    breeding next generation
    generating new random programs

Search complete.  Best score = 10.8166538264

uint16_t gp_rand(uint16_t x) {
    uint16_t r0=0,r1=0,r2=0,r3=0,r4=0;
    uint16_t r5=0,r6=0,r7=0,r8=0,r9=0;

    r2 = r2 * r7;
    r4 = r4 >> (r3 % 16);
    r5 = r5 % r7;
    r1 = r1 - r1;
    r7 = x ^ r7;
    r4 = 3034 + r4;
    r4 = r4 << (r4 % 16);
    r7 = r7 % r4;
    r0 = r0 + r1;
    r0 = 4311 + r0;
    r4 = x * 32200;
```

```
        r9 = x;
        r6 = x + r9;
        r1 = r6 / r1;
        r7 = r7 >> (r2 % 16);
        r2 = r2 | r9;
        r0 = r0 + r4;
        r1 = r1 << (x % 16);
        r1 = r1;
        r2 = 62996 << (63589 % 16);
        r3 = r2;
        r5 = r3 * 2544;
        r1 = r5 + x;
        r0 = r1 + r0;
        return r0;
}
(Runtime: 114.680 seconds)
```

showing that this run found a program that achieved a best score of about 10.82 when compared to a high quality generator. We can use the run.c test program to create an output file,

```
$ ./run 123 100000 gp.out 1
```

which reports the period of this generator to be 65536. We can test the output with ent and with the Monty Hall program we developed in Sect. 1.1 (monty.c). This program will deliver output values of 33.33333% and 66.66666% when supplied with a file of random bytes. Running ent gives,

```
Entropy = 7.999914 bits per byte.

Optimum compression would reduce the size
of this 200000 byte file by 0 percent.

Chi square distribution for 200000 samples is 23.88, and
randomly would exceed this value 99.99 percent of the times.

Arithmetic mean value of data bytes is 127.5200
      (127.5 = random).
Monte Carlo value for Pi is 3.137191372 (error 0.14 percent).
Serial correlation coefficient is -0.001613 (totally
uncorrelated = 0.0).
```

where the entropy is very close to 8. The χ^2 will be unreliable for these small files so we ignore it. The mean byte value is spot on and the estimate of π is close, as is the serial correlation.

The Monty Hall program gives,

```
Results after 100000 simulations:
   Wins on initial guess  = 33286 (33.2860 %)
   Wins on changing guess = 66714 (66.7140 %)
```

which is solidly in the direction we would expect to see. We can visualize the Sierpinski triangle generated by this program, which we used as the metric in the first place, with the sierp16.c program (not listed for space reasons) and the ifs_plot.py script from Sect. 1.4. Doing this gives,

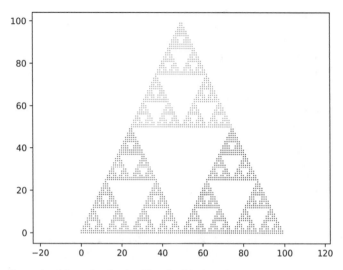

which is discretized but clearly the Sierpinski triangle.

We mentioned above that there is extraneous code in the generated programs. Let's look again at the code for the sample generator,

```
uint16_t gp_rand(uint16_t x) {
    uint16_t r0=0,r1=0,r2=0,r3=0,r4=0;
    uint16_t r5=0,r6=0,r7=0,r8=0,r9=0;

    r2 = r2 * r7;
    r4 = r4 >> (r3 % 16);
    r5 = r5 % r7;
    r1 = r1 - r1;
    r7 = x ^ r7;
    r4 = 3034 + r4;
    r4 = r4 << (r4 % 16);
    r7 = r7 % r4;
    r0 = r0 + r1;
    r0 = 4311 + r0;
    r4 = x * 32200;
    r9 = x;
    r6 = x + r9;
    r1 = r6 / r1;
    r7 = r7 >> (r2 % 16);
    r2 = r2 | r9;
    r0 = r0 + r4;
    r1 = r1 << (x % 16);
    r1 = r1;
    r2 = 62996 << (63589 % 16);
    r3 = r2;
    r5 = r3 * 2544;
    r1 = r5 + x;
    r0 = r1 + r0;
    return r0;
}
```

where we see that the return value is register 0. Working backwards, we can see which instructions are important and which can be eliminated. This gives us the canonical version of the function,

```
uint16_t gp_rand(uint16_t x) {
    uint16_t r0=0,r1=0,r2=0,r3=0,r4=0;
    uint16_t r5=0,r6=0,r7=0,r8=0,r9=0;

    r0 = 4311 + r0;
    r4 = x * 32200;
    r0 = r0 + r4;
    r2 = 62996 << (63589 % 16);
    r5 = r2 * 2544;
    r1 = r5 + x;
    r0 = r1 + r0;
    return r0;
}
```

This program creates the same output files as the evolved function.

Can truly useful pseudorandom number generators be evolved in the first place? We've shown that in our limited 16-bit space we can definitely evolve useful generators. It turns out that this approach works "in the real world", too. In 1991 Koza used GP to evolve a pseudorandom number generator in Lisp [8]. In 2006, Lamenca-Martinez et al. [9] also evolved a generator, again initially in Lisp but translated to C, which did well on DIEHARD tests. Recall that DIEHARD was the precursor to dieharder. Finally, more recently, in 2016 Picek et al. evolved cryptographically secure generators that pass strict randomness tests [10].

7.7 Chapter Summary

In this chapter we examined sequences that exhibit randomness. We looked at the randomness of digits in irrational and transcendental numbers in both base 16 and base 10. We saw that these sequences passed strict randomness tests. We then considered the digits of large factorials and saw that while they show some signs of randomness they are not the best source for the effort involved. Moving on from there, we considered random bits extracted from 1-D elementary cellular automata and demonstrated that at least one of these elementary cellular automata, "Rule 30", does show random behavior. Next, we considered how to use 1-D chaotic maps as a source of randomness and developed a new pseudorandom generator, randlog, based on two logistic maps running in the chaotic region and demonstrated very good performance on the dieharder and TestU01 suites. We also demonstrated how the hallmark of chaotic behavior, sensitive dependence on initial conditions, applies to randlog, even if the starting seeds were separated by as little as 10^{-10} when running parallel streams. The interleaved streams still give nearly perfect results on dieharder. Finally, we had a bit of fun, and developed genetic programming code to evolve 16-bit pseudorandom generators and showed that they performed well for certain tasks.

Exercises

7.1 Using a program like y-cruncher (http://www.numberworld.org/y-cruncher/) or even the Linux desk calculator, dc, calculate many digits of randomly selected irrational square roots and measure how random they are using ent or dieharder. What do you think of the claim that most real numbers are probably normal? (Hint: if using dc run it from the command line:

```
$ dc -e "16 o 100000 k 2 v f" > output
```

which will set the output to base 16, the number of (decimal) digits to 100,000, and then calculate and print the square root of 2 capturing the output in the file "output". Note, the output file will require some further processing.)

7.2 The Rule 30 generator state is always initialized to the same starting value, 0x0000000100000000. Experiment with random state initialization. Does this affect the quality of the output bits? Must the central bit always start in the on position?

7.3 The tent map is a 1-D chaotic map defined for $x_i \in [0, 1]$ as,

$$
x_{i+1} = \begin{cases} ax_i, & 0 \le x_i \le \frac{1}{2} \\ a(1 - x_i), & \frac{1}{2} < x_i \le 1 \end{cases} \tag{7.1}
$$

where $a \in [0, 2]$. If $a = 2 - \epsilon$, the map is in the chaotic region. Is the output of the chaotic region of this map significantly more random than the logistic map for $r = 4 - \epsilon$? Try ϵ values between 10^{-2} and 10^{-5}.

7.4 The evolve.py program detailed in Sect. 7.6 uses the distance between two small Sierpinski triangles as the metric for the quality of the candidate pseudo-random number generator. Incorporate the entropy and estimate of π from ent as well. Does this make a difference regarding the quality and period of the generators found?

7.5 Run the evolve.py program for different top values in $(0, 0.5]$. What effect does changing top (f_k) have on the output quality? Do the same for the population size and maximum program length. To which parameter is the output most sensitive?

References

1. Borel, M. Émile. "Les probabilités dénombrables et leurs applications arithmétiques." Rendiconti del Circolo Matematico di Palermo (1884–1940) 27, no. 1 (1909): 247–271.
2. Kneusel, Ronald T. Numbers and Computers. Springer International Publishing, 2017.
3. Kennedy, R. J. and Eberhart," Particle swarm optimization." In Proceedings of IEEE International Conference on Neural Networks IV, pages, vol. 1000. 1995.
4. Conway, John. "The game of life." Scientific American 223, no. 4 (1970): 4.

5. Wolfram, Stephen. A new kind of science. Vol. 5. Champaign: Wolfram media, 2002.
6. Pareek, Narendra K., Vinod Patidar, and Krishan K. Sud. "A Random Bit Generator Using Chaotic Maps." IJ Network Security 10, no. 1 (2010): 32–38.
7. Koza, John R. Genetic programming: on the programming of computers by means of natural selection. Vol. 1. MIT press, 1992.
8. Koza, John R. "Evolving a Computer Program to Generate Random Numbers Using the Genetic Programming Paradigm." In ICGA, pp. 37–44. 1991.
9. Lamenca-Martinez, Carlos, Julio Cesar Hernandez-Castro, Juan M. Estevez-Tapiador, and Arturo Ribagorda. "Lamar: A new pseudorandom number generator evolved by means of genetic programming." In PPSN, pp. 850–859. 2006.
10. Picek, Stjepan, Dominik Sisejkovic, Vladimir Rozic, Bohan Yang, Domagoj Jakobovic, and Nele Mentens. "Evolving cryptographic pseudorandom number generators." In International Conference on Parallel Problem Solving from Nature, pp. 613–622. Springer International Publishing, 2016.

Index

Printed in the United States
By Bookmasters